Calculus I Exam File

Eric M. Lederer, University of Colorado at Denver and Red Rocks Community College, Editor; D. R. Arterburn, New Mexico Institute of Mining and Technology; Bill Bompart, Augusta College; Peter Braunfeld, University of Illinois at Urbana-Champaign; Frances Burford, Galveston College; William E. Demmon, University of Wisconsin Center - Manitowoc; Mary Elick, Missouri Southern State College; Billy Finch, University of Florida; Michael E. Frantz, University of La Verne; Biswa N. Ghosh, Hudson County College; LeRoy P. Hammerstrom, Eastern Nazarene College; John H. Jenkins, Embry-Riddle Aeronautical University; Anya M. Kroth, West Valley College; Ann F. Landry, Dutchess Community College; David H. Lankford, Bethel College; John Martin, Santa Rosa Junior College; Varoujan Mazmanian, Stevens Institute of Technology; Thomas A. Metzger, University of Pittsburgh; Alejandro Perez, Laredo Junior College; Calvin E. Piston, John Brown University; John Putz, Alma College; Michael Schneider, Belleville Area College; Walter S. Sizer, Moorhead State University; Alan Stickney, Wittenberg University; Joseph F. Stokes, Western Kentucky University; Norman Sweet, State University College; Bill W. Vannatta, Temple Junior College; Robert P. Webber, Longwood College; Joseph E. Wiest, West Virginia Wesleyan College

ENGINEERING PRESS, INC. SAN JOSE, CALIFORNIA

Donald G. Newnan, Ph.D.
EXAM FILE Series Editor

Printed in the United States of America

5 4 3 2 1

Library of Congress Cataloging-in-Publication Data

Calculus I exam file

(Exam file series)
1. Calculus--Problems, exercises, etc. I. Lederer, Eric M., II. Arterburn, D. R. (David R.), III. Title: Calculus 1 exam file.
IV. Series.
QA303.C166 1986 515'.076 86-13535
ISBN 0-910554-61-7

Engineering Press, Inc. P.O. Box 1 San Jose, California 95103-0001

Contents

Foreword

Students studying calculus have traditionally been concerned with preparation for exams. Class notes, textbooks, and textbook supplements have been the usual references in this preparation. All too often they have proved to be slightly inappropriate, because professors actually ask questions that are somewhat different from the ones in these sources. There are various reasons for this, including opinions of what is to be emphasized, awareness of time constraints on students taking tests, and others.

Since students of calculus would like to know on which aspects of the subject they are likely to be tested, the kind of questions that will be asked, and what professors think is a good answer to such questions, this book is designed to satisfy these wants. Many math professors have looked in their exam files and have contributed their favorite calculus exam questions. These questions are given exactly as they appeared in actual course exams, and the solutions are in the professors' own handwriting.

No attempt has been made to impose a uniform format on the problems or the solutions. Thus you will find diverse problem solving styles and notation. This diversity should prove to be one of the real strengths of this book. Each of the contributing professors has been careful not to make errors, but nobody is perfect and it is possible that an error has escaped their attention. If you find one, a note mailed to Engineering Press will be appreciated.

The book is the first of three volumes designed to accompany a standard three course sequence in calculus. We expect that after this book helps you do well in the first course, you will want to use the other two for the later courses.

Eric Lederer
Editor

PRECALCULUS

REAL NUMBERS

■■■-1-1

Solve the inequality $\dfrac{1 - x}{x + 3} \leqslant 0$. Give your answer in interval notation.

****

sign of $1-x$

sign of $x+3$

sign of $\dfrac{1-x}{x+3}$

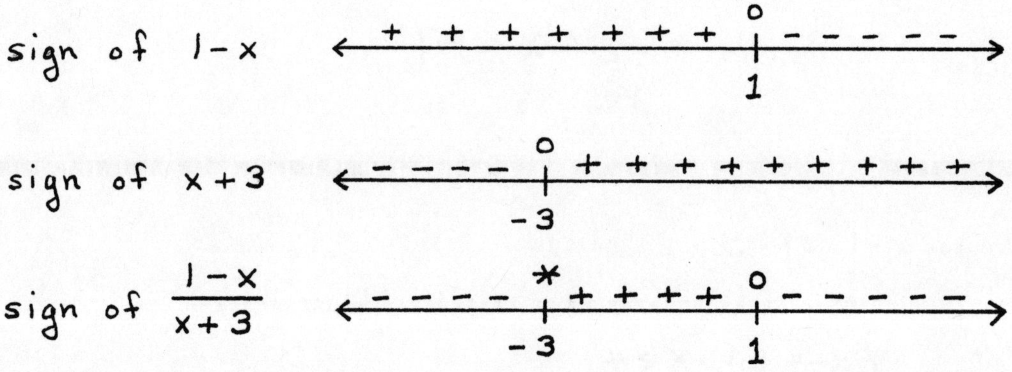

* undefined at $x = -3$

So, the solution is $(-\infty, -3) \cup [1, \infty)$

1-2 ■■■

Solve the inequality $\dfrac{x-3}{x+1} > 5$

****

SOLUTION #1:
$$[x+1>0 \text{ AND } x-3>5(x+1)] \text{ OR } [x+1<0 \text{ AND } x-3<5(x+1)]$$
$$[x>-1 \text{ AND } x<-2] \text{ OR } [x<-1 \text{ AND } x>-2]$$
$$-2<x<-1$$

SOLUTION #2:
$$\frac{x-3}{x+1} - 5 > 0$$
$$\frac{x-3-5(x+1)}{x+1} > 0$$
$$\frac{x+2}{x+1} < 0$$

SOLUTION #3:
$$(x-3)(x+1) > 5(x+1)^2$$
$$(x+2)(x+1) < 0$$

$$
\begin{array}{l}
x+1 \quad -\,-\,-\,-\,-\,-\,-\,|\,-\,-\,|+\,+\,+\,+\,+\,+\,+\,+ \\
x+2 \quad -\,-\,-\,-\,-\,-\,+\,+\,|\,+\,+\,+\,+\,+\,+\,+\,+ \\
\end{array}
$$
$$\quad\quad\quad\quad\quad -2 \quad -1$$

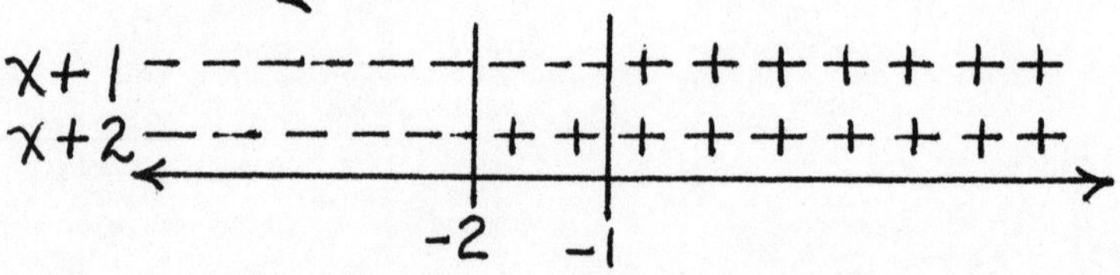

$$-2<x<-1$$

1-3 ■■■

Solve for x : $|x-4| = 2x + 1$

****

$$|x-4| = \begin{cases} x-4 & \text{if } x \geq 4 \\ -(x-4) & \text{if } x < 4 \end{cases}$$

If $x \geq 4$ then $x-4 = 2x+1 \longrightarrow x = -5$ contradiction

If $x < 4$ then $-(x-4) = 2x+1 \longrightarrow x = 1$ $\quad$ ANS: $\underline{x = 1}$

1-4

Find those real numbers which satisfy the following inequality.

$$\frac{|x - 2|}{x + 3} \leq 4$$

****

Since division by zero is prohibited, $x \neq -3$.

Suppose $x < -3$. Then the left side of the inequality is the quotient of a positive number divided by a negative number. Such a negative quotient is less than 4, so at least part of the solution is $x < -3$.

If $-3 < x$ and $x < 2$; then $x - 2 < 0$ so $|x - 2| = 2 - x$, and $0 < x + 3$ so multiplication by $x + 3$ does not reverse the inequality. Then

$$\frac{2 - x}{x + 3} \leq 4$$

$$2 - x \leq 4x + 12$$

$$-10 \leq 5x$$

$$-2 \leq x$$

Then $-2 \leq x < 2$ is also part of the solution.

Now suppose $2 \leq x$. Then the left side of the inequality is a positive number which is less than 1 and so less than or equal to 4.

Therefore the complete solution is

$$x < -3 \quad \text{or} \quad x \geq -2.$$

This may be written

$$(-\infty, -3) \cup [-2, \infty).$$

DISTANCE IN THE COORDINATE PLANE

1-5

Show that the midpoint of the hypotenuse of a right triangle is the center of a circle which contains the vertices of the right triangle.

****

Without loss of generality we will place the right triangle in the following position:

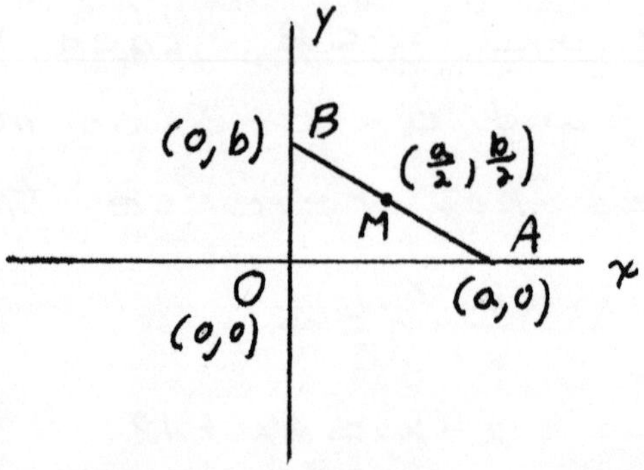

$$\overline{MA} = \sqrt{\left(a - \frac{a}{2}\right)^2 + \left(0 - \frac{b}{2}\right)^2} = \sqrt{\frac{a^2+b^2}{4}} = \frac{\sqrt{a^2+b^2}}{2}$$

$$\overline{MB} = \sqrt{\left(0 - \frac{a}{2}\right)^2 + \left(b - \frac{b}{2}\right)^2} = \sqrt{\frac{a^2+b^2}{4}} = \frac{\sqrt{a^2+b^2}}{2}$$

$$\overline{MO} = \sqrt{\left(\frac{a}{2} - 0\right)^2 + \left(\frac{b}{2} - 0\right)^2} = \sqrt{\frac{a^2+b^2}{4}} = \frac{\sqrt{a^2+b^2}}{2}$$

Since $\overline{MO} = \overline{MA} = \overline{MB}$ we may conclude that M is the center of a circle which contains the vertices of the triangle.

1-6

Prove: The midpoint of the hypotenuse of a right triangle is equidistant from the three vertices.

[Hint: Introduce a coordinate system so that the vertices are located at (0,0),(2a,0) and (0,2b).]

****

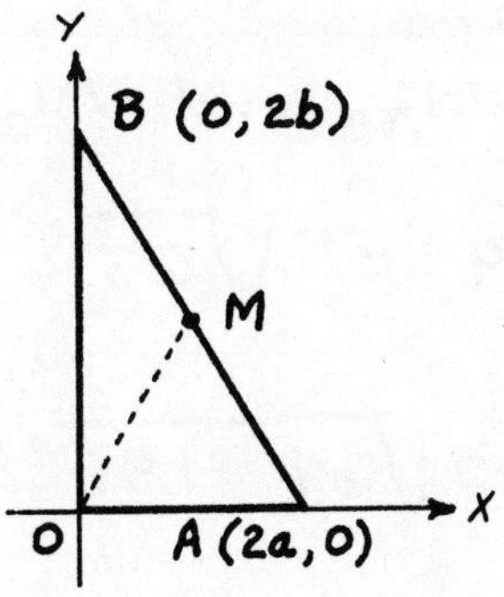

Let the vertices of triangle AOB be as shown in the figure.

If M is the midpoint of $\overline{AB}$ then its coordinates are

$$\left(\frac{2a+0}{2}, \frac{0+2b}{2}\right) = (a, b)$$

Now, the distance from A to M [d(AM)] is given by

$$d(AM) = \sqrt{(2a-a)^2 + (0-b)^2} = \sqrt{a^2+b^2}$$

Since M is the midpoint of AB, d(AM) = d(MB).

And $d(OM) = \sqrt{(0-a)^2 + (0-b)^2} = \sqrt{a^2+b^2}$

So d(AM) = d(MB) = d(OM) as desired.

1-7

Let triangle ABC be defined by the three points A(-4,6), B(8,2) and C(4,-8). Verify that the line segment joining the midpoints of the segments AB and BC is parallel to and one-half the length of line segment AC.

****

M_{AB} = midpoint of segment $\overline{AB} = \left(\frac{-4+8}{2}, \frac{6+2}{2}\right)$

$= (2, 4)$

M_{BC} = midpoint of segment $\overline{BC} = \left(\frac{8+4}{2}, \frac{2-8}{2}\right)$

$= (6, -3)$

$d(M_{AB}, M_{BC})$ = distance between M_{AB} and M_{BC}

$= \sqrt{(2-6)^2 + (4+3)^2} = \sqrt{16+49} = \sqrt{65}$

$d(A, C) = \sqrt{(-4-4)^2 + (6+8)^2} = \sqrt{64+196}$

$= \sqrt{260} = 2\sqrt{65}$

$\therefore\ d(M_{AB}, M_{BC}) = \frac{1}{2}\, d(A, C)$

Also, m = slope of segment $\overline{M_{AB} M_{BC}}$

$= \frac{4+3}{2-6} = -\frac{7}{4}$

m_{AC} = slope of segment $\overline{AC} = \frac{6+8}{-4-4} = -\frac{7}{4}$

$\therefore$ These line segments are parallel, since they have the same slopes.

EQUATION OF A LINE

1-8

Find the equation of the line which is perpendicular to 3y+2x=5 and passing through (-2,5).

****

The given line $3y+2x=5$ can be rewritten as:

$$3y = -2x + 5$$

$$\Rightarrow y = \frac{-2x}{3} + \frac{5}{3}$$

Hence, the slope of the given line $= -\frac{2}{3}$.

Hence, the slope of the line perpendicular to the given line $= +\frac{3}{2}$.

Using the point-slope formula:

$$y - y_0 = m(x - x_0)$$

$$\Rightarrow y - 5 = \frac{3}{2}(x + 2)$$

$$\Rightarrow y - 5 = \frac{3x}{2} + 3$$

$$\Rightarrow y = \frac{3x}{2} + 8$$ is the equation of the required line.

1-9

Find the equation of a line containing the point (4,-1) and perpendicular to the graph of $2x - 3y + 4 = 0$.

****

Any line perpendicular to the graph of $2x - 3y + 4 = 0$ must have the form:

$$3x + 2y + C = 0$$

Since (4,-1) is on this line,

$$3(4) + 2(-1) + C = 0$$

$$C = -10$$

Equation of the line is: $3x + 2y - 10 = 0$.

1-10

Find an equation for the line through the point (-1,2) which is perpendicular to the line $3y + 5 = 7x$.

****

The given line is $y = \frac{7}{3}x - \frac{5}{3}$ so its slope is $7/3$. Therefore, the slope of the desired perpendicular is $-3/7$, and the equation is

$$y - 2 = -\frac{3}{7}(x + 1).$$

1-11

Write the equation of the perpendicular bisector of the segment joining the points (-4,0) and (8,6).

****

THE LINE MUST PASS THROUGH THE MIDPOINT OF THE SEGMENT AND HAVE SLOPE THE NEGATIVE RECIPROCAL OF THE SEGMENT'S SLOPE.

SLOPE OF THE SEGMENT:

$$\frac{6-0}{8-(-4)} = \frac{6}{12} = \frac{1}{2}$$

SLOPE OF A LINE PERPENDICULAR TO THE SEGMENT:

$$-2$$

MIDPOINT OF THE SEGMENT:

$$\left(\frac{-4+8}{2}, \frac{0+6}{2}\right) = (2,3)$$

EQUATION OF THE PERPENDICULAR BISECTOR:

$$y-3 = -2(x-2)$$

$$y-3 = -2x+4$$

$$2x+y = 7$$

1-12

Find the equation of the secant line passing through points (1,-10) and (0,-6) of the graph of $y = x^2 - 5x - 6$.

****

The slope of the line is $m = \frac{-10-(-6)}{1-0} = -4$. Using the formula, $y - y_1 = m(x - x_1)$, we obtain $y - (-10) = -4(x-1)$ or $y = -4x - 6$.

1-13

Line L has equation $3x + 4y = 7$. The point P is (1,2). Find equations of: the line which is parallel to L containing P; the line which is perpendicular to L containing P.

****

$3x + 4y = 7 \Leftrightarrow 4y = -3x + 7 \Leftrightarrow y = -\frac{3}{4}x + \frac{7}{4}$

Slope of L is $m = -3/4$

Parallel line: has slope of $-3/4$, contains (1,2)

Point-slope form is $y - 2 = -\frac{3}{4}(x-1)$

Perpendicular line: has slope of $-1/m = 4/3$, contains (1,2)

Point-slope form is $y - 2 = \frac{4}{3}(x-1)$

1-14

Find the equation of the line which passes through the origin and is perpendicular to the line given by the equation

$$x + 3y = 4$$

****

The slope of the line $3y + x = 4$ OR $y = -1/3\,x + 4/3$ is $-1/3$, thus the slope of a line perpendicular to it has slope of 3.

The equation of any line through the origin is $y = mx$

thus the equation is $\boxed{y = 3x}$

1-15

Find the equation of the perpendicular bisector of the line segment joining (-1, 1) to (3, 4).

****

The midpoint is $\left(\frac{-1+3}{2}, \frac{1+4}{2}\right) = (1, 5/2)$.

The slope of the line given is $\frac{4-1}{3-(-1)} = 3/4$, so a perpendicular line would have slope $-4/3$. Thus the perpendicular bisector is given by $\frac{y-5/2}{x-1} = -4/3$, or

$$y = -\frac{4}{3}x + 23/6 .$$

1-16

Find an equation of the line which passes through the points (-3,2) and (5,-1).

****

$$m = \frac{-1-2}{5+3} = -\frac{3}{8}$$

$$y-2 = -\frac{3}{8}(x+3)$$

$$y = -\frac{3}{8}x + \frac{7}{8}$$

1-17

Find the equation of the line ℓ through the point (1,2) with slope $\frac{1}{4}$.

****

Let (x,y) be any variable point on the line ℓ distinct from the point $(1,2)$. See the diagram below.

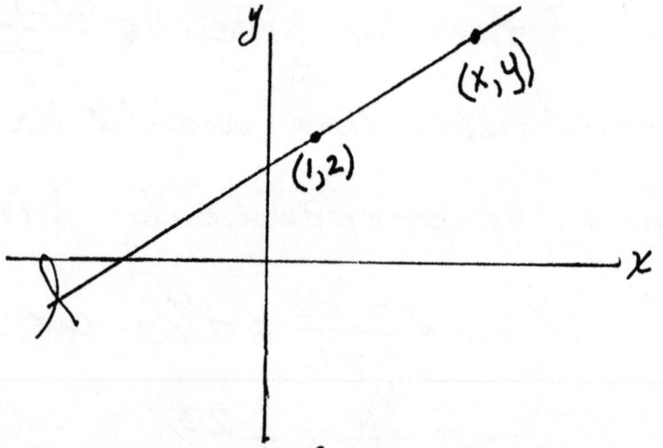

Assume the slope of ℓ is constant.
By the definition of slope $\frac{y-2}{x-1}$ is the slope of ℓ.
Since the slope of ℓ is given to be $\frac{1}{4}$, $\frac{y-2}{x-1} = \frac{1}{4}$.

So $4(y-2) = x-1$

$4y-8 = x-1$

So $-7 = x-4y$

The equation of ℓ is $x-4y=-7$.

FUNCTIONS, RELATIONS, AND GRAPHS

1-18

Determine the domain and range of the function given by

$$f(x) = -\sqrt{1 - \sqrt{x}}\,.$$

****

THE DOMAIN IS THE SET OF ALL x VALUES FOR WHICH THE FUNCTION IS DEFINED. SINCE THE SQUARE ROOT IS DEFINED (AS A REAL NUMBER) ONLY FOR NON-NEGATIVE VALUES,

$$1 - \sqrt{x} \geq 0$$
$$-\sqrt{x} \geq -1$$
$$\sqrt{x} \leq 1$$

SO

$$0 \leq x \leq 1.$$

THEREFORE, THE DOMAIN IS THE SET OF ALL x SUCH THAT $0 \leq x \leq 1$. EXPRESSED AS AN INTERVAL, THE DOMAIN IS $[0, 1]$.

THE RANGE IS THE SET OF ALL VALUES OF THE FUNCTION. BECAUSE THE DOMAIN IS $[0, 1]$, WE HAVE

$$0 \leq x \leq 1$$
$$0 \leq \sqrt{x} \leq 1$$
$$0 \geq -\sqrt{x} \geq -1$$
$$1 \geq 1 - \sqrt{x} \geq 0$$
$$1 \geq \sqrt{1 - \sqrt{x}} \geq 0$$
$$-1 \leq -\sqrt{1 - \sqrt{x}} \leq 0$$
$$-1 \leq f(x) \leq 0.$$

EXPRESSED AS AN INTERVAL, THE RANGE IS $[-1, 0]$.

1-19

Find the range and domain of relation R if $R = \{(x,y)\,|\,x^2 + 6x + y^2 = 4y - 9$ where $x, y \in$ reals$\}$.

****

$$x^2 + 6x + y^2 = 4y - 9$$

$$x^2 + 6x + 9 + y^2 - 4y + 4 = 4$$

$$(x+3)^2 + (y-2)^2 = 2^2$$

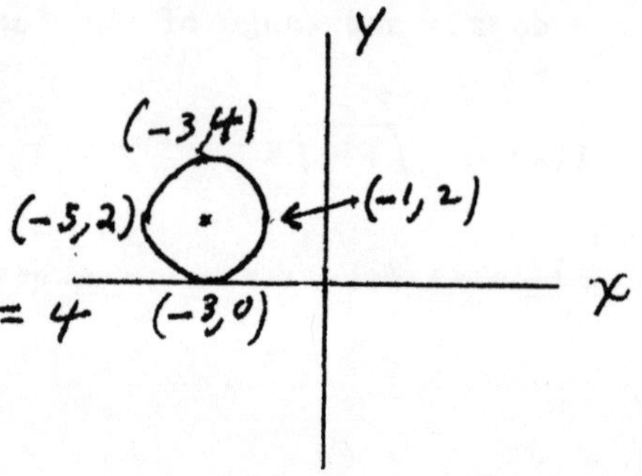

a circle of radius 2 with the center at (-3,2). From the sketch we see that the domain is [-5,-1] and the range is [0,4].

1-20

Sketch $y = \sin^{-1}(\sin x)$.

****

Note that the domain is the reals while the range of $\sin^{-1}$ is $[-\frac{\pi}{2}, \frac{\pi}{2}]$.

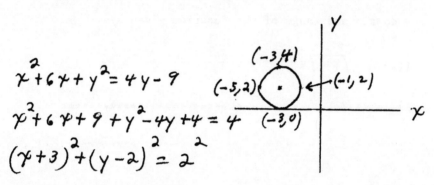

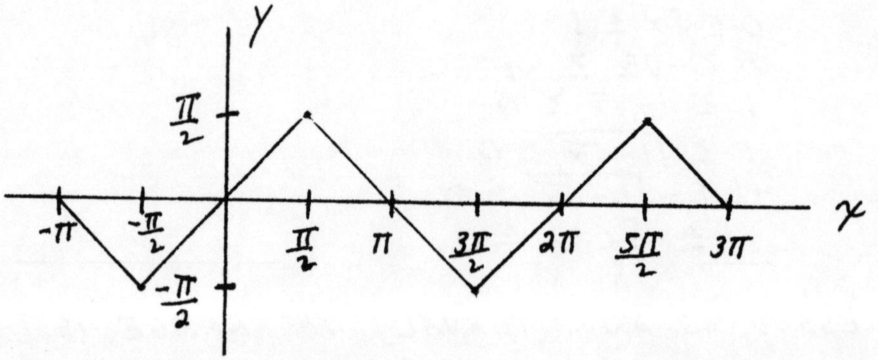

1-21

Let A be the set containing the elements 1, 2, and 0. Let B be the set of real numbers. Let f be the function from A to B such that for each x in A, f(x)= 3x+1. Sketch a graph of f(x).

****

We first find f(1), f(2) and f(0) from the definition of the given function, $f(x) = 3x+1$.

$f(1) = 3(1)+1 = 4$, $f(2) = 3(2)+1 = 7$, $f(0) = 3(0)+1 = 1$

Since the graph of f is the geometric picture of all points (x, y) such that $y = f(x) = 3x+1$, we need to locate the points $(1,4)$, $(2,7)$ and $(0,1)$. So the graph of $f(x) = 3x+1$ is given below.

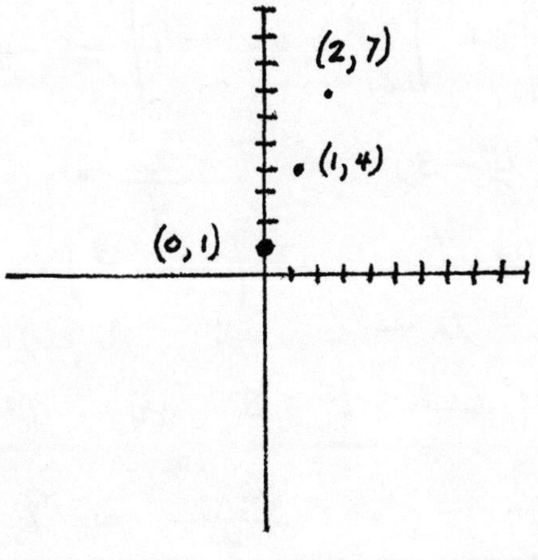

1-22

Show that the product of two odd functions is an even function.

****

Let $f(x)$ and $g(x)$ be odd functions and $h(x) = f(x)g(x)$. Now

$$\begin{aligned} h(-x) &= f(-x)g(-x) \\ &= \{-f(x)\}\{-g(x)\} \\ &= f(x)g(x) \\ &= h(x). \end{aligned}$$

Since $h(-x) = h(x)$ we may conclude that $h(x)$ is an even function and the statement follows.

MISCELLANEOUS PROBLEMS ABOUT FUNCTIONS

1-23

Determine a and b so that the equation $x^2 + y^2 + ax - 6y + 4 = 0$ represents a circle with center at (2,b) and radius 3.

****

Starting with $x^2+y^2+ax-6y+4=0$ and completing squares we obtain

$$\left[x^2+ax+\left(\frac{a}{2}\right)^2\right] + \left[y^2-6y+9\right] = -4+\left(\frac{a}{2}\right)^2+9$$

or $\left(x+\frac{a}{2}\right)^2 + (y-3)^2 = 5+\frac{a^2}{4}$. This is a circle with center at $\left(-\frac{a}{2}, 3\right)$. In order for the center to be at $(2, b)$ it must be that $-\frac{a}{2} = 2$ and $b = 3$ or $a = -4$ and $b=3$.

Now $R^2 = 5+\frac{a^2}{4} = 5+\frac{(-4)^2}{4} = 9$, verifying that the radius is 3.

1-24

Graph the function $y = \sqrt{1 - 4x^2}$. Using this, graph

$$y = 1 + \sqrt{1 - 4(x - 2)^2}.$$

****

The original function is

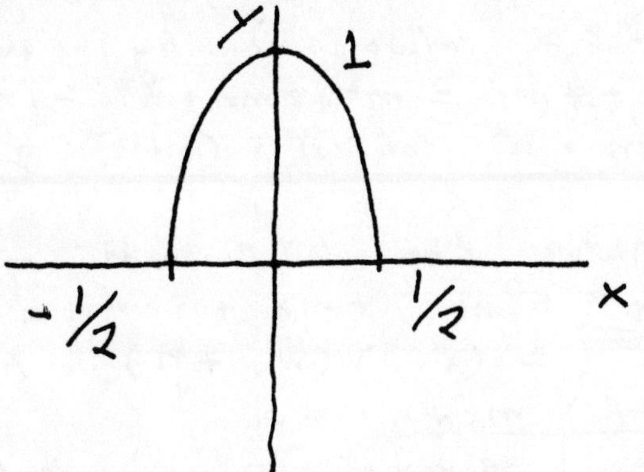

,

so the shifted function is

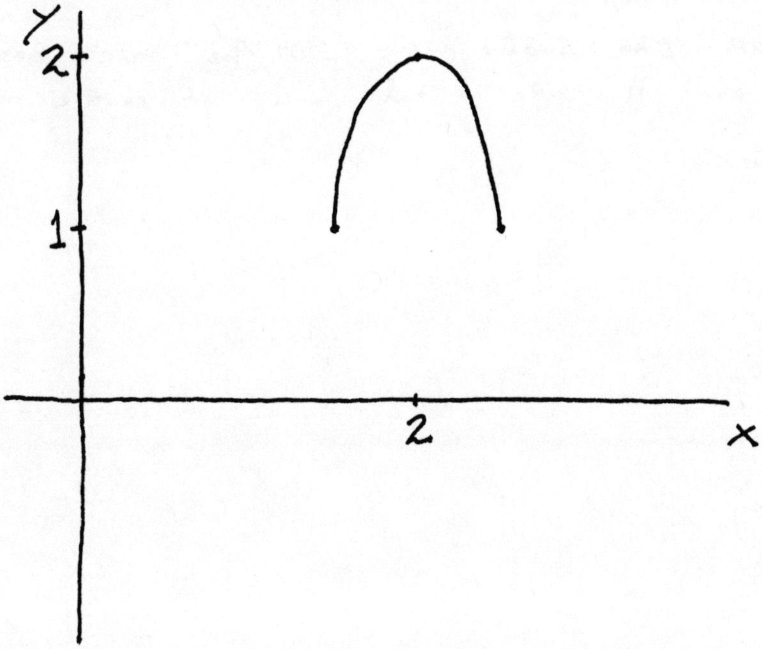

1-25

Show that the equation $mx^2 + (m+n)x + n = 0$ always has real roots.

****

METHOD 1: By comparing with the standard form $Ax^2 + Bx + C = 0$, $A = m$, $B = m+n$ and $C = n$. Recall that the Discriminant $D = B^2 - 4AC$. (Will show that $D \geq 0$ always.) By substitution $D = (m+n)^2 - 4mn = m^2 + 2mn + n^2 - 4mn$. Hence $D = m^2 - 2mn + n^2$ or $D = (m-n)^2 \geq 0$ ∎

METHOD 2: FACTOR the left side as follows:

$$mx^2 + (m+n)x + n = mx^2 + mx + nx + n = mx(x+1) + n(x+1)$$
$$= (x+1)(mx+n). \text{ Hence}$$

$$(x+1)(mx+n) = 0$$

and $x+1=0$ OR $mx+n=0$ leads to

$x = -1$ OR $x = -\dfrac{n}{m}$

NOTE: For the Root $x = -n/m$ we assume that m and n are Real and of course $m \neq 0$.

2 LIMITS AND CONTINUITY

THE LIMIT OF A FUNCTION

2-1

Find $\lim_{x \to 4} \dfrac{x^2 - x - 12}{x^2 - 16}$ if possible.

****

$\lim_{x \to 4} (x^2 - x - 12) = 16 - 4 - 12 = 0$ and

$\lim_{x \to 4} (x^2 - 16) = 16 - 16 = 0$. Therefore

$\lim_{x \to 4} \dfrac{x^2 - x - 12}{x^2 - 16} = \dfrac{0}{0}$ which is indeterminant.

Factoring the numerator and denominator and canceling common factors produces

$$\lim_{x \to 4} \frac{(x-4)(x+3)}{(x-4)(x+4)} = \lim_{x \to 4} \frac{x+3}{x+4} = \frac{4+3}{4+4} = 7/8$$

2-2

Evaluate the following limit:

$$\lim_{x \to 0} \sqrt{\frac{\tan 9x}{\tan x}}$$

****

$$\lim_{x \to 0} \sqrt{\frac{\tan 9x}{\tan x}} = \lim_{x \to 0} \sqrt{\frac{\frac{\sin 9x}{\cos 9x}}{\frac{\sin x}{\cos x}}} = \lim_{x \to 0} \sqrt{\frac{\sin 9x}{\sin x} \cdot \frac{\cos x}{\cos 9x}}$$

$$= \sqrt{\lim_{x \to 0} \frac{\sin 9x}{\sin x} \cdot \frac{\cos x}{\cos 9x}}$$

$$= \sqrt{\lim_{x \to 0} \frac{\sin 9x}{\sin x} \cdot \lim_{x \to 0} \frac{\cos x}{\cos 9x}}$$

$$\lim_{x \to 0} \frac{\cos x}{\cos 9x} = \frac{1}{1} = 1 \qquad \lim_{x \to 0} \frac{\sin 9x}{\sin x} = \lim_{x \to 0} \frac{\sin 9x}{9x} \cdot \frac{9x}{\sin x}$$

$$= 9 \lim_{x \to 0} \frac{\sin 9x}{9x} \cdot \lim_{x \to 0} \frac{1}{\frac{\sin x}{x}}$$

Consider the unit circle with θ in radians:

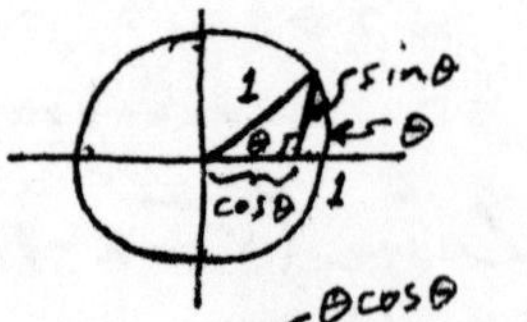

$$\theta \cos \theta \le \sin \theta \le \theta$$

$$\cos \theta \le \frac{\sin \theta}{\theta} \le 1$$

$$\lim_{\theta \to 0} : \quad 1 \le \frac{\sin \theta}{\theta} \le 1$$

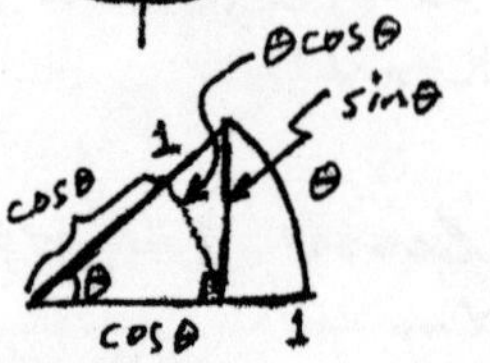

Therefore $\lim_{\theta \to 0} \frac{\sin \theta}{\theta} = 1 = \frac{\sin m\theta}{m\theta}$

Then, $\lim_{x \to 0} \frac{\sin 9x}{\sin x} = 9 \cdot 1 \cdot 1 = 9$

$$\lim_{x \to 0} \sqrt{\frac{\tan 9x}{\tan x}} = \sqrt{9 \cdot 1} = 3$$

$$\therefore \lim_{x \to 0} \sqrt{\frac{\tan 9x}{\tan x}} = 3.$$

2-3

Find each limit if it exists:

a. $\lim_{x \to 5} (2x^2 + \sqrt{3x + 1})$

b. $\lim_{x \to -2} \frac{x^2 - x - 6}{x + 2}$

c. $\lim_{x \to 9} \frac{\sqrt{x} - 3}{x - 9}$

d. $\lim_{x \to 4} (3x + \sqrt{x - 4})$

****

a. Since there are no denominators or any other things to make this complicated, this is found by substitution thus:

$$\lim_{x \to 5} (2x^2 + \sqrt{3x+1}) = 2 \cdot 5^2 + \sqrt{3 \cdot 5 + 1} = 50 + 4 = 54$$

b. $$\lim_{x \to -2} \frac{(x+2)(x-3)}{x+2} = \lim_{x \to -2} (x-3) = -2 - 3 = -5$$

c. $$\lim_{x \to 9} \frac{\sqrt{x} - 3}{x - 9} \cdot \frac{\sqrt{x} + 3}{\sqrt{x} + 3}$$ (introducing the conjugate of the numerator)

$$= \lim_{x \to 9} \frac{x - 9}{(x-9)(\sqrt{x}+3)}$$

$$= \lim_{x \to 9} \frac{1}{\sqrt{x} + 3} = \frac{1}{\sqrt{9} + 3} = \frac{1}{6}$$

d. This limit <u>does not exist</u> because the value can not be calculated for any values of x that are less than 4.

2-4

Prove, using the definition of a limit, that $\lim_{x \to 2} (4x+3) = 11$.

****

We must show that for all $\varepsilon > 0$ there exists $\delta > 0$ such that $|(4x+3)-11| < \varepsilon$ whenever $|x-2| < \delta$. That is, if we specify an $\varepsilon > 0$ we want to be able to find a corresponding $\delta > 0$ such that $|x-2| < \delta$ forces $|(4x+3)-11| < \varepsilon$.

$|4x+3-11| = |4x-8| = |4(x-2)| = 4|x-2| < \varepsilon$, which is equivalent to $|x-2| < \varepsilon/4$. Let $\delta = \varepsilon/4$.

Given $\varepsilon > 0$, by letting $\delta = \varepsilon/4$ it must be true that $|4x+3-11| < \varepsilon$ whenever $|x-2| < \delta$. Since we were able to find the δ that works, it is true that

$$\lim_{x \to 2} (4x+3) = 11 .$$

2-5

Evaluate $\lim_{x \to 1} \dfrac{x^2 - 5x + 4}{x^2 + 2x - 3}$.

****

$$\lim_{x \to 1} \frac{x^2-5x+4}{x^2+2x-3} = \lim_{x \to 1} \frac{(x-4)\cancel{(x-1)}}{(x+3)\cancel{(x-1)}} = \frac{1-4}{1+3} = -\frac{3}{4} .$$

2-6

Evaluate the limits (a) $\lim_{x \to 1} \dfrac{1 - \frac{1}{x^3}}{1 - \frac{1}{x^2}}$ (b) $\lim_{x \to 2} \dfrac{\sqrt{x} - \sqrt{2}}{x^2 - 4}$

****

(a) $\lim_{x \to 1} \dfrac{1 - \frac{1}{x^3}}{1 - \frac{1}{x^2}} = \dfrac{1-1}{1-1} = \dfrac{0}{0}$ notice evaluating this limit produces an indeterminate result.

We need to change the problem algebraically.

$= \lim_{x \to 1} \dfrac{\frac{x^3 - 1}{x^3}}{\frac{x^2 - 1}{x^2}}$ combine top and bottom into single fractions

$$= \lim_{x \to 1} \frac{x^3-1}{x^3} \cdot \frac{x^2}{x^2-1} = \lim_{x \to 1} \frac{(x-1)(x^2+x+1) \cdot x^2}{x^3(x-1)(x+1)} = \lim_{x \to 1} \frac{(x^2+x+1)}{x(x+1)} = \frac{3}{2}$$

(b) $\lim_{x \to 2} \dfrac{\sqrt{x} - \sqrt{2}}{x^2 - 4} = \dfrac{\sqrt{2} - \sqrt{2}}{4-4}$ again indeterminate.

This time rationalize the top line.

$$= \lim_{x \to 2} \frac{\sqrt{x} - \sqrt{2}}{x^2-4} \cdot \frac{\sqrt{x}+\sqrt{2}}{\sqrt{x}+\sqrt{2}} = \lim_{x \to 2} \frac{x-2}{(x^2-4)(\sqrt{x}+\sqrt{2})} = \lim_{x \to 2} \frac{x-2}{(x-2)(x+2)(\sqrt{x}+\sqrt{2})}$$

$$= \lim_{x \to 2} \frac{1}{(x+2)(\sqrt{x}+\sqrt{2})} = \frac{1}{4(\sqrt{2}+\sqrt{2})} = \frac{1}{4 \cdot 2\sqrt{2}} = \frac{1}{8\sqrt{2}}$$

2-7

Find the limit of $f(x) = 2-|x|$ as x approaches zero using your intuition and then prove your result is correct by the δ, ε definition of limit.

****

The limit $L = 2$. The Proof follows:

Given any $\epsilon > 0$ we must find a $\delta > 0$ such that whenever $0 < |x-a| < \delta$ then $|f(x) - L| < \epsilon$ i.e.
" $0 < |x| < \delta$ then $|2 - |x| - 2| < \epsilon$ using $a = 0$, $f(x) = 2 - |x|$ and $L = 2$.

Now $|f(x) - L| < \epsilon \Leftrightarrow |2 - |x| - 2| < \epsilon \Leftrightarrow |-|x|| < \epsilon$ ($\Leftrightarrow$ Reads "is equivalent to")
Recall that $|-a| = |a|$ so $|-|x|| = |x|$. Hence

$$|f(x) - L| < \epsilon \Leftrightarrow |x| < \epsilon \text{ and choose } \delta = \epsilon.$$

Retracing our steps yields,
$$0 < |x| < \delta \Rightarrow |x| < \epsilon \Rightarrow |-|x|| < \epsilon \Rightarrow |\underbrace{2 - |x|}_{f(x)} - \underbrace{2}_{L}| < \epsilon \blacksquare$$

2-8

Prove that $\lim_{x \to 4} \left(\frac{x}{2} - 3\right) = -1$

****

LET $\epsilon > 0$. THEN WE SEEK $\delta > 0$ SUCH THAT IF $0 < |x - 4| < \delta$ THEN $\left|\frac{x}{2} - 3 + 1\right| < \epsilon$

$$\left|\frac{x}{2} - 3 + 1\right| = \frac{1}{2}|x - 4|$$

CHOOSE $\delta = 2\epsilon$

THEN, $|x - 4| < \delta \Rightarrow |x - 4| < 2\epsilon$

$$\Rightarrow \frac{|x-4|}{2} < \epsilon$$

$$\Rightarrow \left|\frac{x}{2} - 3 + 1\right| < \epsilon$$

2-9

For the function $f(x) = x^2 + 2x - 1$ evaluate

$$\lim_{h \to 0} \frac{f(x+h) - f(x)}{h}$$

****

$$f(x) = x^2 + 2x - 1$$

$$f(x+h) = (x+h)^2 + 2(x+h) - 1$$

$$= x^2 + 2xh + h^2 + 2x + 2h - 1$$

$$f(x+h) - f(x) = 2xh + h^2 + 2h$$

thus $\lim_{h \to 0} \frac{f(x+h) - f(x)}{h}$

$$= \lim_{h \to 0} \frac{2xh + h^2 + 2h}{h}$$

$$= \lim_{h \to 0} (2x + h + 2) = \boxed{2x + 2}$$

ONE-SIDED LIMITS

2-10

Find the limit of F at x = 1 if it exists, where

$$F(x) = \begin{cases} x^2 - 4, & x \geq 1 \\ 2x - 3, & x < 1 \end{cases}$$

****

$$\lim_{x \to 1^+} F(x) = \lim_{x \to 1^+} (x^2 - 4) = (1)^2 - 4 = -3$$

$$\lim_{x \to 1^-} F(x) = \lim_{x \to 1^-} (2x - 3) = 2(1) - 3 = -1$$

Since $\lim_{x \to 1^+} F(x) \neq \lim_{x \to 1^-} F(x)$, the limit of F at $x = 1$ does not exist.

2-11

Find the limit of G at x = -1 if it exists, where

$$G(x) = \begin{cases} x^2 - x - 1, & x < -1 \\ 3 + 2x, & x \geq -1 \end{cases}$$

****

$$\lim_{x \to -1^+} G(x) = \lim_{x \to -1^+} (3 + 2x) = 3 + 2(-1) = 1$$

$$\lim_{x \to -1^-} G(x) = \lim_{x \to -1^-} (x^2 - x - 1) = (-1)^2 - (-1) - 1 = 1$$

Since $\lim_{x \to -1^+} G(x) = \lim_{x \to -1^-} G(x) = 1$, the limit of G at $x = -1$ is 1.

2-12

Determine whether $\lim_{x \to 0} \frac{1}{1 + 5^{\frac{1}{x}}}$ exists.

****

$$\lim_{x \to 0^+} \frac{1}{1+5^{\frac{1}{x}}} = 0 \quad , \quad \lim_{x \to 0^-} \frac{1}{1+5^{\frac{1}{x}}} = 1$$

$\therefore \lim_{x \to 0} \frac{1}{1+5^{\frac{1}{x}}}$ DOES NOT EXISTS BECAUSE THE ONE-SIDED LIMITS ARE NOT EQUAL.

2-13

Find $\lim_{x \to 2^-} \left(\dfrac{x^2 - x - 3}{x^2 - 3x + 2} \right)$

****

substituting $x = 2$ gives "$\frac{-1}{0}$" which is infinite, so limit is ∞ or $-\infty$.

If x is near 2 on the left (e.g. 1.99),

$$\frac{x^2 - x - 3}{x^2 - 3x + 2} = \frac{x^2 - x - 3}{(x-2)(x-1)}$$

$$= \frac{(\text{near } -1)}{(\text{negative near } 0)(\text{near } 1)} > 0.$$

Therefore, the limit is $+\infty$.

2-14

Evaluate the following limit:

$$\lim_{x \to 3^+} \frac{x^2 + x - 6}{18 - 3x - x^2}$$

****

$$\lim_{x \to 3^+} \frac{x^2+x-6}{18-3x-x^2} = \lim_{x \to 3^+} \frac{x^2+x-6}{(6+x)(3-x)}$$

$$= -\infty$$

since $\lim_{x \to 3^+} x^2+x-6 = 6$,

$\lim_{x \to 3^+} 6+x = 9$, and $\lim_{x \to 3^+} 3-x = 0^-$.

2-15

Sketch a graph of the piecewise defined function f(x). Then use your graph to evaluate each limit or function value question.

$$f(x) = \begin{cases} x, & x \geq 3 \\ x^2, & 0 < x < 3 \\ 1/x, & -1 < x < 0 \\ -1, & x \leq -1 \end{cases}$$

$\lim_{x \to 0^+} f(x) =$ $\quad$ $\lim_{x \to 3^-} f(x) =$

$\lim_{x \to -\infty} f(x) =$ $\quad$ $\lim_{x \to +\infty} f(x) =$

$\lim_{x \to 2} f(x) =$ $\quad$ $f(3) =$

****

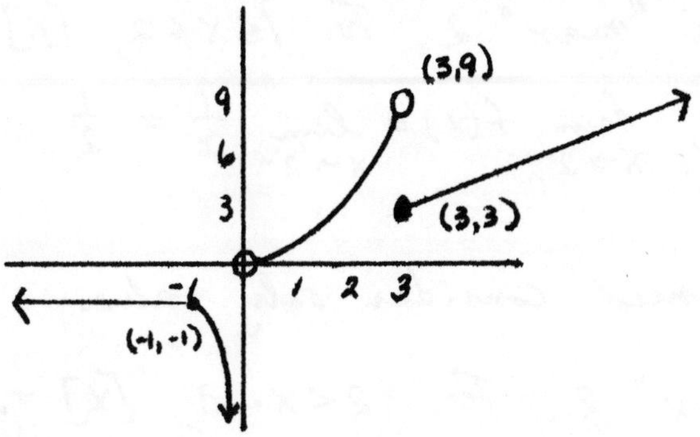

Evaluate the limits by inspection of the graph.

$\lim_{x \to 0^+} (f(x)) = 0$ as x gets near 0 from the right, y gets near 0.

$\lim_{x \to 3^-} f(x) = 9$ as x gets near 3 from the left, y gets near 9

$\lim_{x \to -\infty} f(x) = -1$ as x becomes negatively unbounded, y has the value -1.

$\lim_{x \to +\infty} f(x) = +\infty$ as x becomes positively unbounded, y also becomes positively unbounded.

$\lim_{x \to 2} f(x) = 4$ as x gets near 2 from the left and right, y gets near 4.

$f(3) = 3$ notice, the point (3,9) is not included in the graph.

2-16

Let [x] = greatest integer function. Let $f(x) = \frac{[x]}{x}$. Find each of the following limits, if they exist

(a) $\lim_{x\to 2^-} f(x)$ (b) $\lim_{x\to 2^+} f(x)$ (c) $\lim_{x\to 2} f(x)$

****

(a) to find $\lim_{x\to 2^-} f(x)$ we need consider only values of $x < 2$ and "near" 2. For $1 \le x < 2$, $[x] = 1$, so $f(x) = \frac{1}{x}$. Hence, $\lim_{x\to 2^-} f(x) = \lim_{x\to 2^-} \frac{1}{x} = \frac{1}{2}$

(b) For $\lim_{x\to 2^+} f(x)$ we need consider only values of $x > 2$, but "near" 2. For $2 < x < 3$, $[x] = 2$, so $f(x) = \frac{2}{x}$. Hence $\lim_{x\to 2^+} f(x) = \lim_{x\to 2^+} \frac{2}{x} = 1$

(c) From (a), (b) above: $\lim_{x\to 2^-} f(x) = \frac{1}{2}$, while $\lim_{x\to 2^+} f(x) = 1$. the two one-sided limits are different, so $\lim_{x\to 2} f(x)$ does not exist.

2-17

Calculate: $\lim_{x\to 1} \frac{x^2 + |x-1| - 1}{|x-1|}$

****

$$\lim_{x\to 1^+} \frac{x^2+x-1-1}{x-1} = \lim_{x\to 1^+} \frac{x^2+x-2}{x-1} = \lim_{x\to 1^+} \frac{(x+2)\cancel{(x-1)}}{\cancel{(x-1)}} = 3$$

$$\lim_{x\to 1^-} \frac{x^2-(x-1)-1}{-(x-1)} = \lim_{x\to 1^-} \frac{x^2-x}{-(x-1)} = \lim_{x\to 1^-} \frac{x\cancel{(x-1)}}{-\cancel{(x-1)}} = -1$$

$\lim_{x\to 1} \frac{x^2+|x-1|-1}{|x-1|}$ does not exist because the right hand limit does not equal the left hand limit.

2-18

Evaluate the given limit:

$$\lim_{x\to 3^-} \sqrt{|x|-x}$$

****

AS $x \to 3^-$, $|x| = x$ (FOR $x > 0$), SO

$$\lim_{x\to 3^-} \sqrt{|x|-x} = \lim_{x\to 3^-} \sqrt{x-x} = \lim_{x\to 3^-} 0 = 0$$

(NOTE: DO NOT CONFUSE $x \to 3^-$ AND $x \to -3$)

LIMITS AT INFINITY

2-19

Find $\lim_{x \to +\infty} \frac{4x^2 + x}{x^2 - 2}$ if possible.

****

Dividing numerator and denominator by x^2, which is the highest power of x present produces

$$\lim_{x \to +\infty} \frac{\frac{4x^2}{x^2} + \frac{x}{x^2}}{\frac{x^2}{x^2} - \frac{2}{x^2}} = \lim_{x \to +\infty} \frac{4 + \frac{1}{x}}{1 - \frac{2}{x^2}}$$

But since $\lim_{x \to +\infty} \frac{1}{x} = 0$ and $\lim_{x \to +\infty} \frac{2}{x^2} = 0$ it follows that

$$\lim_{x \to +\infty} \frac{4x^2 + x}{x^2 - 2} = \frac{4 + 0}{1 - 0} = 4.$$

2-20

Find the limit of $x - \sqrt{x^2 - 4}$ as x approaches positive infinity.

****

$$\lim_{x \to +\infty} (x - \sqrt{x^2 - 4}) = \lim_{x \to +\infty} (x - \sqrt{x^2 - 4}) \cdot \frac{(x + \sqrt{x^2 - 4})}{(x + \sqrt{x^2 - 4})}$$

$$= \lim_{x \to +\infty} \frac{x^2 - (x^2 - 4)}{x + \sqrt{x^2 - 4}} = \lim_{x \to +\infty} \frac{4}{x + \sqrt{x^2 - 4}} = 0.$$

2-21

Use algebraic techniques to find:

$$\lim_{x \to +\infty} (x - \sqrt{x^2 - 3x})$$

****

Since $x \to +\infty$ it is helpful to manipulate the function so that powers of x become powers of $1/x$. Begin by rationalizing the numerator:

$$\lim_{x \to +\infty} (x - \sqrt{x^2 - 3x}) = \lim_{x \to +\infty} \left[\frac{x - \sqrt{x^2 - 3x}}{1} \cdot \frac{x + \sqrt{x^2 - 3x}}{x + \sqrt{x^2 - 3x}} \right]$$

$$= \lim_{x \to +\infty} \left[\frac{x^2 - (x^2 - 3x)}{x + \sqrt{x^2 - 3x}} \right]$$

Now divide both numerator and denominator by x noting that as $x \to +\infty$, $\sqrt{x^2} = |x| = x$.

$$= \lim_{x \to +\infty} \left[\frac{3x}{x + \sqrt{x^2 - 3x}} \cdot \frac{1/x}{1/\sqrt{x^2}} \right]$$

$$= \lim_{x \to +\infty} \left[\frac{3}{1 + \sqrt{1 - 3/x}} \right]$$

$$= \frac{3}{2}$$

Since $3/x \to 0$ as $x \to +\infty$.

2-22

$$\lim_{x\to\infty} \frac{3x^2+4x+2}{5-x+x^3} =$$

****

$$\frac{3x^2+4x+2}{x^3-x+5} = \frac{x^2(3+4/x+2/x^2)}{x^3(1-1/x^2+5/x^3)}$$

$$= \frac{(3+4/x+2/x^2)}{x(1-1/x^2+5/x^3)} \quad \text{If } x \geq 1 \text{ SAY.}$$

THUS

$$\lim_{x\to\infty} \frac{3x^2+4x+2}{5-x+x^3} = \lim_{x\to\infty} \frac{(3+4/x+2/x^2)}{x(1-1/x^2+5/x^3)} = 0.$$

CONTINUITY OF FUNCTIONS

2-23

Tell where f(x) is continuous if $f(x) = \frac{\sqrt{x+7}}{x^2-1}$.

****

Since f(x) is algebraic it will be continuous wherever it is defined. For f(x) to be defined we need

(1) $x+7 \geq 0$, so $x \geq -7$, and

(2) $x^2-1 \neq 0$, so $x \neq \pm 1$.

f(x) is continuous for $-7 \leq x < -1$, $-1 < x < 1$, and $x > 1$.

2-24

Identify any discontinuities in $f(x) = \dfrac{1 - x^4}{x - x^3}$

and classify them as removable or non-removable.

****

$$f(x) = \frac{1-x^4}{x-x^3}$$

factor the expression

$$f(x) = \frac{(1-x^2)(1+x^2)}{x(1-x^2)} = \frac{(1-x)(1+x)(1+x^2)}{x(1-x)(1+x)}$$

We have discontinuities when the denominator = 0.
Thus we have discontinuities at $x=0$, $x=1$ and $x=-1$
A discontinuity is removable at $x=a$ if the limit exists at $x=a$.
Consider the limit at each point

(a) $\displaystyle\lim_{x\to 0} \frac{(1-x)(1+x)(1+x^2)}{x(1-x)(1+x)} = \lim_{x\to 0} \frac{1+x^2}{x} = \frac{1}{0}$ does not exist

(b) $\displaystyle\lim_{x\to 1} \frac{(1-x)(1+x)(1+x^2)}{x(1-x)(1+x)} = \lim_{x\to 1} \frac{1+x^2}{x} = \frac{2}{1} = 2$

(c) $\displaystyle\lim_{x\to -1} \frac{(1-x)(1+x)(1+x^2)}{x(1-x)(1+x)} = \lim_{x\to -1} \frac{1+x^2}{x} = \frac{2}{-1} = -2$

Since $\lim_{x\to 0} f(x)$ does not exist we have a non-removable discontinuity at $x=0$. However, $\lim_{x\to 1} f(x)$ exists and $\lim_{x\to -1} f(x)$ exists, so there are removable discontinuities at $x=1, x=-1$

2-25

Determine whether the following function is continuous at x = 3

$$f(x) = \begin{cases} \dfrac{3x - 2}{7} & x \geq 3 \\ \dfrac{x^2-5x+6}{x-3} & x < 3 \end{cases}$$

****

(i) Is this function defined at $x = 3$?

$$\frac{3(3) - 2}{7} = \frac{9-2}{7} = \frac{7}{7} = 1. \quad \text{(Yes!)}$$

(ii) Does $\lim_{x \to 3} f(x)$ exist?

$$\lim_{x \to 3^+} f(x) = \lim_{x \to 3^+} \frac{3x-2}{7} = \frac{7}{7} = 1$$

$$\lim_{x \to 3^-} f(x) = \lim_{x \to 3^-} \frac{x^2-5x+6}{x-3}$$

$$= \lim_{x \to 3^-} \frac{(x-2)\cancel{(x-3)}}{\cancel{x-3}} = 1$$

Since the right-handed limit equals the left-handed limit,

$$\Rightarrow \lim_{x \to 3} f(x) = 1.$$

(iii) $\lim_{x \to 3} f(x) = f(3) = 1.$

Since the limit as $x \to 3$ equals $f(3)$; the function is continuous at $x = 3$.

2-26

Define a function f(x) by:

$$f(x) = \begin{cases} \dfrac{2x^2 - 9x - 5}{10 + 13x - 3x^2} & , x \neq 5,\ x \neq -2/3 \\ 0 & , x = -2/3 \\ 11/17 & , x = 5 \end{cases}$$

Find all values of x for which this function is discontinuous. For each discontinuity, determine if it is a removable discontinuity or an essential discontinuity. If removable, explain how.

****

$$\lim_{x\to 5} \frac{2x^2-9x-5}{10+13x-3x^2} = \lim_{x\to 5} \frac{(2x+1)(x-5)}{(5-x)(2+3x)}$$

$$= \lim_{x\to 5} -\frac{(2x+1)(5-x)}{(5-x)(2+3x)} = \lim_{x\to 5} -\frac{2x+1}{2+3x} = -\frac{11}{17}$$

$\therefore$ f(x) has a <u>removable</u> discontinuity at $x=5$, since $\lim\limits_{x\to 5} f(x) \neq f(5)$. To remove the discontinuity, define $f(5) = -11/17$.

$$\lim_{x\to -2/3} f(x) = \lim_{x\to -2/3} -\frac{2x+1}{2+3x}$$ and this limit clearly does not exist, since the numerator approaches $-1/3$ and the denominator approaches 0. Hence, f(x) has an <u>essential</u> discontinuity at $x = -2/3$.

2-27

Sketch a graph of <u>one</u> function which satisfies all of the following properties:

i) $f(0) = 4$

ii) $\lim_{x \to 0} f(x)$ does not exist

iii) $\lim_{x \to 3^+} f(x) = -2$

iv) $\lim_{x \to 3^-} f(x) = 2$

v) $\lim_{x \to 5} f(x) = 4$

vi) $f(3) = 0$

****

Note : there is no unique solution: One possible graph is given. The behavior of the function around x=0, x=3 and x=5 is critical.

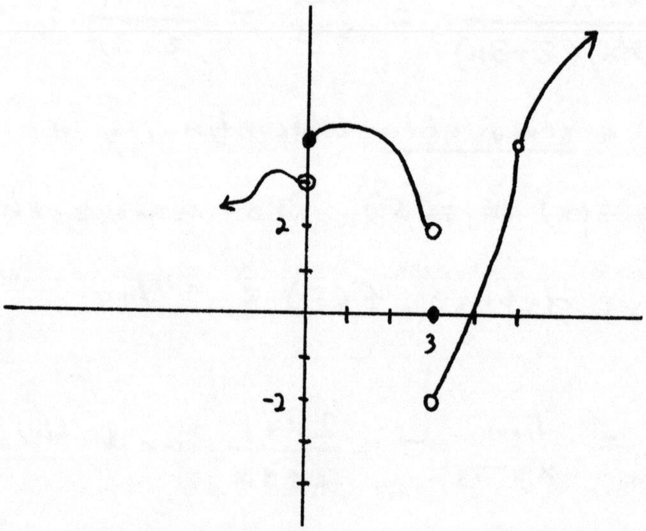

2-28

For $f(x) = \dfrac{x - 3}{x^2 - 9}$ name any points of discontinuity. Describe what is happening to the graph of f(x) at each point of discontinuity.

****

f(x) will be discontinuous at all points where the denominator goes to zero $x^2 - 9 = 0 \Rightarrow x = \pm 3$

$$f(x) = \frac{x-3}{(x+3)(x-3)} = \frac{1}{x+3}, \quad x \neq \pm 3$$

at $x = 3$: since the factor $x-3$ reduces away, the point $x=3$ is a missing point discontinuity represented by a "hole" in the graph at the point where $x=3$.

at $x = -3$: since the factor $x+3$ does not reduce away, the point $x=-3$ is a point of infinite discontinuity represented in the graph by a vertical asymptote along the line $x = -3$.

2-29

Determine whether f(x) is continuous at c:

$$f(x)=\begin{cases} 2-5x^2 & \text{if } x<0 \\ 2 & \text{if } x=0 \\ \sqrt{4-x^2} & \text{if } 0<x<2 \end{cases} \qquad c=0$$

****

NOTE: f(x) IS CONTINUOUS AT C IF (1) f(c) IS DEFINED, (2) $\lim_{x\to c} f(x)$ EXISTS, AND (3) $\lim_{x\to c} f(x) = f(c)$

(1) $f(0) = 2$

(2) WE SHOW $\lim_{x\to 0} (x)$ EXISTS BY SHOWING THAT $\lim_{x\to 0^+} f(x) = 2$ AND $\lim_{x\to 0^-} f(x) = 2$:

$$\lim_{x\to 0^+} f(x) = \lim_{x\to 0^+} \sqrt{4-x^2} = \sqrt{\lim_{x\to 0^+}(4-x^2)} = \sqrt{4} = 2$$

$$\lim_{x\to 0^-} f(x) = \lim_{x\to 0^-}(2-5x^2) = 2-5\cdot 0 = 2$$

$$\therefore \lim_{x\to 0} f(x) = 2$$

(3) $\lim_{x\to 0} f(x) \overset{?}{=} (0)$ YES, BECAUSE $f(0)=2$ AND $\lim_{x\to 0} f(x) = 2$

CONCLUSION: f(x) IS CONTINUOUS AT C=0

2-30

Given the function f defined by

$$f(x) = \begin{cases} \dfrac{x^2 + 2x - 8}{x - 2} & \text{for } x \neq 2 \\ N & \text{for } x = 2, \end{cases}$$

either find a value of N which makes f continuous or show that no value of N can make f continuous.

****

IN ORDER THAT f BE CONTINUOUS AT 2, WE MUST HAVE

$$\lim_{x \to 2} f(x) = f(2).$$

BUT $f(2) = N$, SO WE MUST HAVE

$$\lim_{x \to 2} f(x) = N.$$

NOW $\lim_{x \to 2} f(x) = \lim_{x \to 2} \dfrac{x^2+2x-8}{x-2}$ SINCE $x \neq 2$ AS $x \to 2$

$$= \lim_{x \to 2} \frac{(x-2)(x+4)}{x-2}$$

$= \lim_{x \to 2} (x+4)$ SINCE $x \neq 2$ AS $x \to 2$

$$= 6.$$

THEREFORE, $N = 6$ MAKES f CONTINUOUS.

2-31

Sketch the graph of $f(x) = \begin{cases} 0, & x \le -2 \\ (x+2)^2, & -2 < x < 0 \\ x+4, & 0 < x < 1 \\ 3, & x \ge 1 \end{cases}$ and find all discontinuities.

****

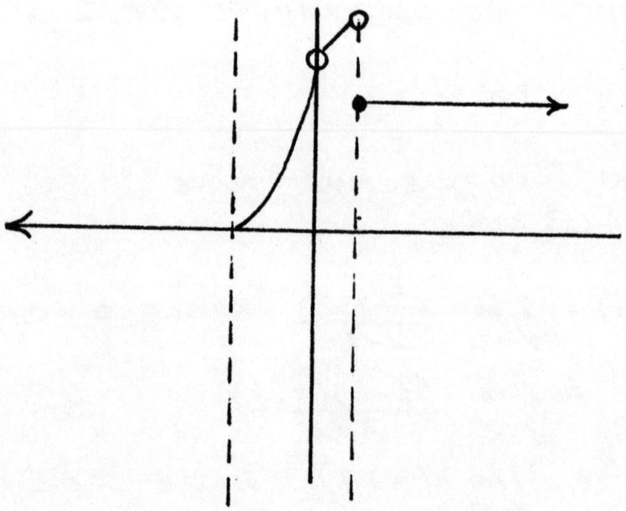

DISCONTINUOUS AT

$x = 0$ ($f(0)$ UNDEFINED)

$x = 1$ ($\lim_{x\to 1} f(x)$ DOES NOT EXIST)

LIMITS AND ASYMPTOTES

■■ **2-32**

Find all the vertical and horizontal asymptotes for the graph

of the function $f(x) = \dfrac{x^2 - 4}{9 - x^2}$.

****

$f(x) = \dfrac{x^2 - 4}{(3-x)(3+x)}$ is infinite at $x = \pm 3$.

So, vertical asymptotes are $x = 3$ and $x = -3$.

$\lim\limits_{x \to \infty} \dfrac{x^2 - 4}{9 - x^2} = \lim\limits_{x \to \infty} \dfrac{1 - \frac{4}{x^2}}{\frac{9}{x^2} - 1} = -1$.

So, $y = -1$ is a horizontal asymptote.

■■ **2-33**

Find $\lim\limits_{x \to +\infty} \dfrac{x^2 - x^3}{x^3}$.

****

$\lim\limits_{x \to +\infty} \dfrac{x^2 - x^3}{x^3} = \lim\limits_{x \to +\infty} \dfrac{x^2}{x^3} - \lim\limits_{x \to +\infty} \dfrac{x^3}{x^3}$

$= 0 - 1$

$= -1$

2-34

Find the equation of the oblique asymptote that

$$y = \frac{4x^3 + 6x^2 - 10x + 6}{2x^2 + 4}$$

approaches as x approaches positive infinity.

****

$$y = \frac{4x^3 + 6x^2 - 10x + 6}{2x^2 + 4} = \frac{2x^3 + 3x^2 - 5x + 3}{x^2 + 2}.$$

$$\begin{array}{r|l}
 & 2x + 3 \\
x^2 + 2 & 2x^3 + 3x^2 - 5x + 3 \\
 & 2x^3 \qquad\quad + 4x \\
\hline
 & \quad 3x^2 - 9x + 3 \\
 & \quad 3x^2 \qquad + 6 \\
\hline
 & \qquad -9x - 3
\end{array}$$

Now we have that $y = 2x + 3 - \dfrac{9x+3}{x^2+2}$.

Clearly as $x \to \infty$ we see that $y \to 2x + 3$ $\therefore$ $y = 2x + 3$ is the oblique asymptote.

2-35

Find all asymptotes (both vertical and horizontal) for the graph of the function

$$f(x) = \frac{x^2 - 16}{x^2 - 5x + 4}.$$

****

$$f(x) = \frac{(x-4)(x+4)}{(x-4)(x-1)}$$

VERTICAL ASYMPTOTES:

(THE LINE $x=a$ IS A VERT. ASYMP. IF $\lim_{x \to a} |f(x)| = \infty$.)

THE ONLY POSSIBILITIES ARE $x=4$ AND $x=1$.

$$\lim_{x \to 4} \left| \frac{(x-4)(x+4)}{(x-4)(x-1)} \right| = \lim_{x \to 4} \left| \frac{x+4}{x-1} \right| \text{ SINCE } x \neq 4 \text{ AS } x \to 4$$
$$= 8/3.$$

SO $x=4$ IS NOT A VERT. ASYMP.

$$\lim_{x \to 1} \left| \frac{(x-4)(x+4)}{(x-4)(x-1)} \right| = \lim_{x \to 1} \left| \frac{x+4}{x-1} \right| \text{ SINCE } x \neq 4 \text{ FOR } x \text{ NEAR } 1$$
$$= \infty.$$

SO $x=1$ IS A VERT. ASYMP.

HORIZONTAL ASYMPTOTES:

$$\lim_{x \to \infty} \frac{(x-4)(x+4)}{(x-4)(x-1)} = \lim_{x \to \infty} \frac{x+4}{x-1} = \lim_{x \to \infty} \frac{1+\frac{4}{x}}{1-\frac{1}{x}} = 1$$

SO $y=1$ IS A HORIZONTAL ASYMPTOTE IN THE POSITIVE DIRECTION.

$$\lim_{x \to -\infty} \frac{(x-4)(x+4)}{(x-4)(x-1)} = \lim_{x \to -\infty} \frac{x+4}{x-1} = \lim_{x \to -\infty} \frac{1+4/x}{1-1/x} = 1$$

SO $y=1$ IS A HORIZONTAL ASYMPTOTE IN THE NEGATIVE DIRECTION.

2-36

Find the vertical and horizontal asymptotes of the function

$$f(x) = \frac{3x^2 + 100}{4x^2 - 100}$$

****

For the vertical asymptotes, set the denominator equal to 0:

$$4x^2 - 100 = 0, \text{ so } x^2 = 25 \text{ and } x = \pm 5$$

Vertical asymptotes: lines $x = 5$ and $x = -5$

For the horizontal asymptotes, we need

$$\lim_{x \to +\infty} f(x) \quad \text{and} \quad \lim_{x \to -\infty} f(x)$$

$$\lim_{x \to +\infty} \frac{3x^2 + 100}{4x^2 - 100} = \lim_{x \to +\infty} \frac{3 + 100/x^2}{4 - 100/x^2} = \frac{3}{4}$$

Likewise, $\lim_{x \to -\infty} f(x) = 3/4$

Hence, the horizontal asymptote is: $y = 3/4$

2-37

Consider the graph of f where

$$f(x) = \frac{(x+1)(x-1)}{2x}$$

a. Find the asymptotes. b. Show why f is odd or even. c. Sketch.

****

a. $f(x) = \frac{x^2-1}{2x} = \frac{x}{2} - \frac{1}{2x}$. As x approaches zero, f(x) approaches $\pm\infty$, so the vertical axis is an asymptote. As x approaches $\pm\infty$, f(x) approaches $\frac{x}{2}$, so the line $y = \frac{1}{2}x$ is an asymptote.

b. $f(-x) = \frac{(-x)^2-1}{2\cdot(-x)} = \frac{x^2-1}{-2x} = -\frac{x^2-1}{2x}$.

Since $f(-x) = -f(x)$, f is odd.

c.

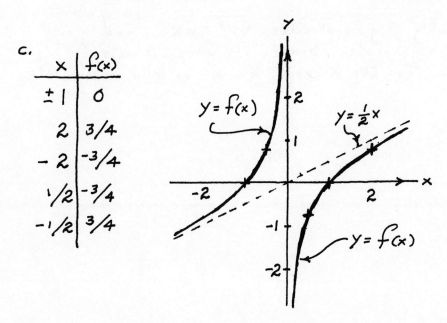

x	f(x)
±1	0
2	3/4
−2	−3/4
1/2	−3/4
−1/2	3/4

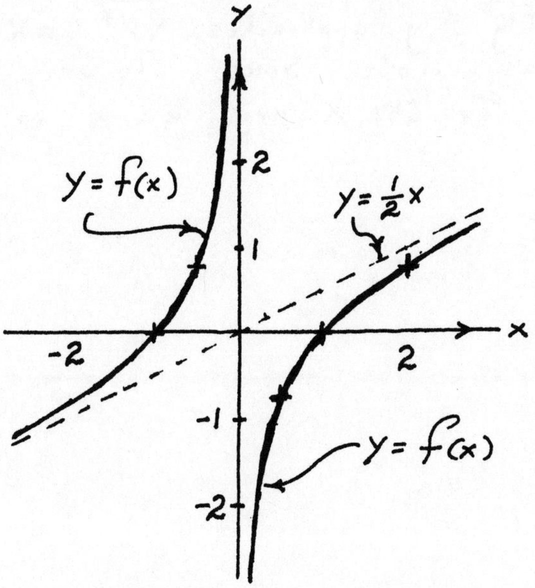

2-38

Show that the hyperbola $x^2-y^2 = 1$ has $y = \pm x$ as asymptotic lines.

****

Begin with $x^2 - y^2 = 1$ and solve for y.

$y^2 = x^2 - 1$. Since the graph is symmetric to the y axis we select the principle root,

$y = \sqrt{x^2-1}$. Now rewrite the radicand as follows:

$$x^2-1 = (x^2-1)\cdot\frac{x^2}{x^2} = \frac{(x^2-1)}{x^2}\cdot x^2 = \left(1-\frac{1}{x^2}\right)\cdot x^2.$$ Hence

as x "gets large" $\frac{1}{x^2}$ approaches zero. By substitution $y = \sqrt{\left(1-\frac{1}{x^2}\right)x^2} = \sqrt{x^2}\sqrt{1-\frac{1}{x^2}}$ and as $x \to \infty$

$\sqrt{1-\frac{1}{x^2}}$ approaches $\sqrt{1} = 1$ by the above.

Consequently, y approaches $\sqrt{x^2} = x$ and $y = x$ is an asymptote. Since the graph is also symmetric to the x axis $y = -x$ is also an asymptote.

2-39

Find all asymptotes for the graph of the function $f(x) = \dfrac{x^3 - 2x + 3}{x^2 + 4x + 4}$.

****

Vertical Asymptotes: $x^2+4x+4 = (x+2)^2 = 0$ at $x = -2$

$\therefore$ one vertical asymptote at $x = -2$

Horizontal Asymptote: $\lim_{x \to +\infty} f(x) = +\infty$

$\therefore$ no horizontal asymptote

Oblique or other asymptote:

$$x^2+4x+4 \overline{\big)\, x^3 \qquad -2x+3} \quad = x - 4 + \frac{10x+19}{x^2+4x+4}$$

$$\begin{array}{r} x^3 + 4x^2 + 4x \\ \hline -4x^2 - 6x + 3 \\ -4x^2 - 16x - 16 \\ \hline 10x + 19 \end{array}$$

$$\lim_{x \to +\infty} \frac{10x+19}{x^2+4x+4} = 0$$

$\therefore$ one oblique asymptote at $y = x - 4$.

2-40

Find the limit of $f(x) = \frac{x^3+x^2-x-1}{x-1}$ as a) x approaches zero and b) x approaches one.

****

a) $\lim_{x \to 0} f(x) = \lim_{x \to 0} \frac{x^3+x^2-x-1}{x-1} = \frac{\lim_{x \to 0} (x^3+x^2-x-1)}{\lim_{x \to 0} (x-1)}$

Hence $\lim_{x \to 0} f(x) = \frac{-1}{-1} = 1$.

b) Since $\lim_{x \to 1} (x-1) = 0$ the quotient Rule fails.

However, $f(x) = \frac{x^2(x+1)-(x+1)}{x-1} = \frac{(x+1)(x^2-1)}{x-1} = (x+1)^2$

The last equality holds whenever $x \neq 1$ and this is tailor-made for the limit, since x never equals one anyway. We now proceed with the limit. $\lim_{x \to 1} f(x) = \lim_{x \to 1} (x+1)^2 = 4$.

3
THE DERIVATIVE

THE DERIVATIVE OF A REAL FUNCTION

3-1

Let $f(x) = 3x - 2x^2$. Use the definition of the derivative as the limit of a difference quotient to find $f'(x)$.

****

$$\text{Diff. Quot.} = \frac{(3(x+h) - 2(x+h)^2) - (3x - 2x^2)}{h}$$

$$= \left(\frac{1}{h}\right)(3x + 3h - 2x^2 - 4xh - 2h^2 - 3x + 2x^2)$$

$$= \left(\frac{1}{h}\right)(3h - 4xh - 2h^2) = 3 - 4x - 2h$$

$$f'(x) = \lim_{h \to 0} (3 - 4x - 2h) = 3 - 4x .$$

3-2

Find the derivative of $f(x) = x^2 - 3$ from the definition.

****

$$f'(x) = \lim_{h \to 0} \frac{f(x+h) - f(x)}{h}$$

$$= \lim_{h \to 0} \frac{(x+h)^2 - 3 - (x^2 - 3)}{h}$$

$$= \lim_{h \to 0} \frac{2xh + h^2}{h}$$

$$= \lim_{h \to 0} (2x + h)$$

$$= 2x$$

3-3

Use the definition of the derivative to find the slope of the tangent to the curve $y = x^2 - 3$ at the point (1,-2).

****

Definition of the derivative:

$$f'(x) = \lim_{h \to 0} \frac{f(x+h) - f(x)}{h}$$

In this case, $f'(x) = \lim_{h \to 0} \frac{[(x+h)^2 - 3] - [x^2 - 3]}{h}$

$$= \lim_{h \to 0} \frac{x^2 + 2xh + h^2 - 3 - x^2 + 3}{h}$$

$$= \lim_{h \to 0} (2x + h) = 2x$$

At point (1,-2), $m = f'(1) = 2(1) = 2$.

3-4

a) Explain how a function could fail to be differentiable at a point.

b) Sketch the graphs of three different functions which fail to be differentiable at a point.

****

a) Since the derivative of a function f is the function f' defined by

$$f'(x) = \lim_{h \to 0} \frac{f(x+h) - f(x)}{h}$$

a function would fail to be differentiable at any point where this limit fails to exist.

b) There are three common situations where a function fails to be differentiable.

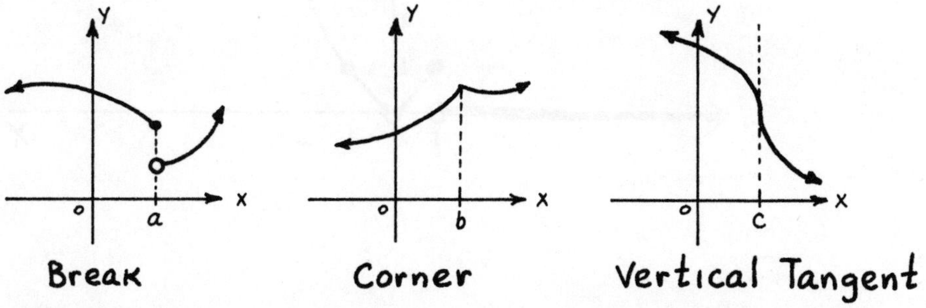

In each case it should be clear that the double sided limit from part (a) does not exist.

3-5

Let f(x) be defined as given below.

(a) Sketch the graph of the function.

(b) For what values of x is the function discontinuous?

(c) Does the derivative of the function exists at x = 1? Why or why not?

$$f(x) = \begin{cases} 0 & \text{if } x < -1 \\ -x & \text{if } -1 \le x < 0 \\ x & \text{if } 0 < x \le 1 \\ x^2 & \text{if } x > 1 \end{cases}$$

****

(a)

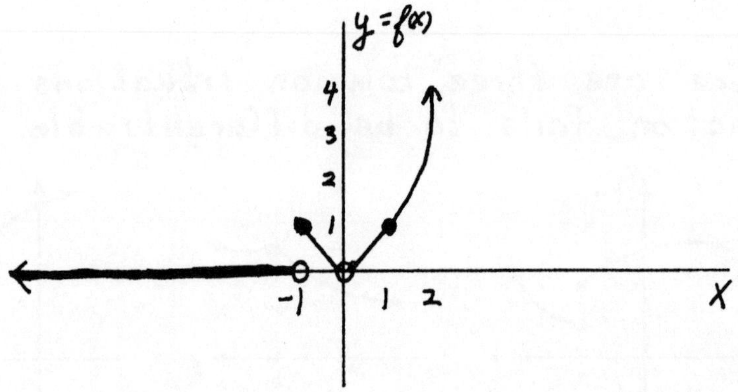

(b) AT −1 BECAUSE $\lim_{x \to -1} f(x)$ DOES NOT EXISTS.

AT 0 BECAUSE f(x) IS NOT DEFINED AT X = 0.

(c) NO, BECAUSE

$$\lim_{h \to 0^+} \frac{f(1+h)-f(1)}{h} = \lim_{h \to 0^+} \frac{(1+h)^2-1^2}{h} = \lim_{h \to 0^+} \frac{2h+h^2}{h} = \lim_{h \to 0^+} (2+h) = 2$$

$$\lim_{h \to 0^-} \frac{f(1+h)-f(1)}{h} = \lim_{h \to 0^-} \frac{1+h-1}{h} = \lim_{h \to 0^-} \frac{h}{h} = \lim_{h \to 0^-} 1 = 1$$

BASIC RULES FOR DERIVATIVES

3-6

Given $f(3) = 5$, $f'(3) = 1.1$, $g(3) = -4$, and $g'(3) = 0.7$, find the following values:

(a) $(f+g)'(3)$

(b) $(f-g)'(3)$

(c) $(f\cdot g)'(3)$

(d) $(f/g)'(3)$

****

(a) $(f+g)'(3) = f'(3) + g'(3) = 1.1 + 0.7 = 1.8$

(b) $(f-g)'(3) = f'(3) - g'(3) = 1.1 - 0.7 = 0.4$

(c) $(f\cdot g)'(3) = f'(3)g(3) + f(3)g'(3) = (1.1)(-4) + (5)(0.7)$

$= -4.4 + 3.5 = -0.9$

(d) $$\left(\frac{f}{g}\right)'(3) = \frac{f'(3)g(3) - f(3)g'(3)}{(g(3))^2} = \frac{(1.1)(-4) - (5)(0.7)}{(-4)^2}$$

$$= \frac{-4.4 - 3.5}{16} = \frac{-7.9}{16} = -0.49375$$

3-7

Given the function $f(x) = x-1 + |x|$, find the value of the derivative at a) $x = 1$, and b) $x = 0$. Hint: The sum rule fails at $x = 0$.

****

a) Sub. $u = |x| = \begin{cases} x, & x \geq 0 \\ -x, & x < 0 \end{cases}$ Then

$f(x) = x-1+u$ and $f'(x) = 1 + u'$. Clearly $u' = 1$ at $x=1$ $\therefore$ $f'(1) = 1+1 = 2$

b) $u(x)$ is not differentiable at $x=0$, but by definition $f'(0) = \lim_{x\to 0} \frac{f(x)-f(0)}{x-0} = \lim_{x\to 0} \frac{x-1+|x|+1}{x}$

Hence $f'(0) = \lim_{x\to 0} \frac{x+|x|}{x}$. EXAMINE Left and Right hand limits.

RT.: $\lim_{x\to 0^+} \frac{x+|x|}{x} = \lim_{x\to 0^+} \frac{2x}{x} = 2$

(See def. of $u(x)$ in part a).)

LT.: $\lim_{x\to 0^-} \frac{x+|x|}{x} = \lim_{x\to 0^-} (0) = 0$.

Hence no limit exists and $f(x)$ is not differentiable at $x=0$.

3-8

Find dy/dx for $y = \dfrac{3x^4 - 2x^2 + x}{3x^2}$

****

Method 1: As a quotient

$$\frac{dy}{dx} = \frac{(3x^2)(12x^3 - 4x + 1) - (3x^4 - 2x^2 + x)(6x)}{(3x^2)^2}$$

$$= \frac{18x^5 - 3x^2}{9x^4} = \frac{6x^3 - 1}{3x^2}$$

Method 2: Simplify first

$$y = \frac{3x^4}{3x^2} - \frac{2x^2}{3x^2} + \frac{x}{3x^2} = x^2 - 2/3 + (1/3)x^{-1}$$

$$\frac{dy}{dx} = 2x - (1/3)x^{-2} = 2x - \frac{1}{3x^2} = \frac{6x^3 - 1}{3x^2}$$

3-9

If u and v are functions of x, find the second derivative of uv with respect to x.

****

$$\frac{d^2}{dx^2}(uv) = \frac{d}{dx}\left[\frac{d}{dx}(uv)\right] = \frac{d}{dx}\left[u\frac{dv}{dx} + v\frac{du}{dx}\right]$$

$$= u\frac{d^2v}{dx^2} + \frac{du}{dx}\frac{dv}{dx} + v\frac{d^2u}{dx^2} + \frac{dv}{dx}\frac{du}{dx}$$

$$= u\frac{d^2v}{dx^2} + 2\frac{du}{dx}\frac{dv}{dx} + v\frac{d^2u}{dx^2}$$

3-10

Given the function $f(x) = x\,|x-1|$, find the value of the derivative at a) $x = o$ and b) $x = 1$. Hint: The product rule fails at $x = 1$.

****

a)

SUB. $u = |x-1| = \begin{cases} x-1, & x \geq 1 \\ 1-x, & x < 1 \end{cases}$ Then

$f(x) = x \cdot u$ and $f'(x) = x \cdot u' + u$ Now u' is the slope of $u(x)$ and is clearly -1 at $x=0$ and $f'(0) = 0(-1) + |0-1| = 1$

b) $u(x)$ is not differentiable at $x=1$, but by definition $f'(1) = \lim\limits_{x\to 1} \dfrac{f(x) - f(1)}{x-1} = \lim\limits_{x\to 1} \dfrac{x|x-1| - 0}{x-1}$.

Examine left and Right hand limits.

RT: $\lim\limits_{x\to 1^+} \dfrac{x|x-1|}{x-1} = \lim\limits_{x\to 1^+} x\left(\dfrac{x-1}{x-1}\right) = 1(1) = 1$

(see def. of $u(x)$ in part a).)

LT.: $\lim\limits_{x\to 1^-} \dfrac{x|x-1|}{x-1} = \lim\limits_{x\to 1^-} x\left(\dfrac{1-x}{x-1}\right) = (1)(-1) = -1$

Hence no limit exists and $f(x)$ is not differentiable at $x = 1$.

THE TANGENT LINE

3-11

a) Find an equation of the line which is tangent to the graph of $f(x) = -x^4 + 2x^2 + x$ at the point $(1,2)$.

b) Determine if the line in part (a) is tangent to f at a different point.

****

a) If $f(x) = -x^4 + 2x^2 + x$ then $f'(x) = -4x^3 + 4x + 1$ and $f'(1) = 1$.

An equation of the line containing the point $(1,2)$ and having a slope of 1 is:

$$y - 2 = 1(x - 1) \quad \text{or} \quad \underline{\underline{y = x + 1}}$$

b) If $y = x + 1$ is tangent to f at a different point, say (a, b), then $f'(a) = 1$.

But $f'(a) = 1 \Rightarrow -4a^3 + 4a + 1 = 1$
$\Rightarrow -4a(a^2 - 1) = 0$
$\Rightarrow a = 0$ or $a = \pm 1$

Now $a = 1 \Rightarrow b = 2$, the point from part (a).
$a = 0 \Rightarrow b = 0$, not on the line $y = x + 1$
$a = -1 \Rightarrow b = 0$, which is on $y = x + 1$

So $y = x + 1$ is tangent to the graph of $f(x)$ at $(1,2)$ and at $\underline{\underline{(-1,0)}}$.

3-12

Find an equation of the line tangent to $f(x) = x^2 - 4x$ at the point (3,-3).

****

$f'(x) = 2x - 4$

$\therefore m = f'(3) = 2(3) - 4 = 2$ is the slope.

$y - y_0 = m(x - x_0) \quad \therefore y + 3 = 2(x - 3)$

OR $\quad 2x - y - 9 = 0$

3-13

Find the equation of a line with slope ½ tangent to the curve $y = 3X^3$.

****

On the given curve $y' = 9x^2$, so setting $9x^2 = \frac{1}{2}$, we obtain $x^2 = 1/18$, $x = \frac{1}{\pm 3\sqrt{2}}$. The corresponding y-values are $\frac{\pm 1}{18\sqrt{2}}$. Thus there are two such lines,

$$\frac{y - \frac{1}{18\sqrt{2}}}{x - \frac{1}{3\sqrt{2}}} = \frac{1}{2} \quad \left(\text{or } y = \frac{1}{2}x - \frac{1}{9\sqrt{2}}\right)$$

and

$$\frac{y - \left(\frac{-1}{18\sqrt{2}}\right)}{x - \left(\frac{-1}{3\sqrt{2}}\right)} = \frac{1}{2} \quad \left(\text{or } y = \frac{1}{2}x + \frac{1}{9\sqrt{2}}\right).$$

3-14

FIND THE EQUATION OF THE LINE WHICH IS TANGENT TO THE GRAPH OF THE FUNCTION $f(x) = 3x^2 - 3x$ and WHICH IS PARALLEL TO THE X- AXIS.

****

$y = 3x^2 - 3x$

$y' = 6x - 3$. For the line to be parallel to the x-axis the slope [hence the derivative at the point of tangency] must be 0.

To find the point where this occurs, we have

$6x - 3 = 0$

$x = 1/2$

$y = 3(1/2)^2 - 3(1/2) = -3/4$

Since the slope is 0 and the line passes through $(1/2, -3/4)$ the equation of the line is

$$\boxed{y = -3/4}$$

3-15

Find an equation of the line tangent to the curve $y = x + 1/x$ at the point $(5, 26/5)$.

****

Slope of tangent line is $m = y' = 1 - \frac{1}{x^2}$, thus the slope at the point $(5, 26/5)$ is $m = 1 - \frac{1}{5^2} = 1 - \frac{1}{25} = 24/25$

Point-slope form of the line is $y - \frac{26}{5} = \frac{24}{25}(x - 5)$

3-16

Discuss the tangent line to the function $f(x) = |x-1|$ at the point (1,0).

****

By definition $f(x) = \begin{cases} x-1 & , x \geq 1 \\ 1-x & , x < 1 \end{cases}$

The graph follows:

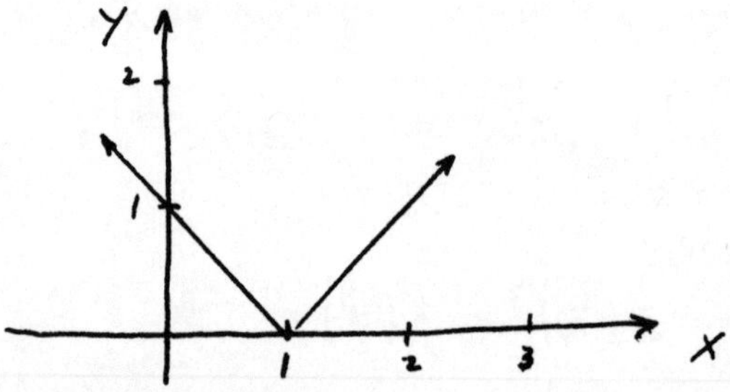

By def. the __tangent line__ to a curve at a pt. P is the limiting position of the secant lines PQ as $Q \to P$ along the curve. Here pt. P is (1,0).

The Right-handed student would probably select the secant lines as follows:

Clearly all the secant lines are already coincident with the line $y = x-1$ and the student would conclude this as the tangent line.

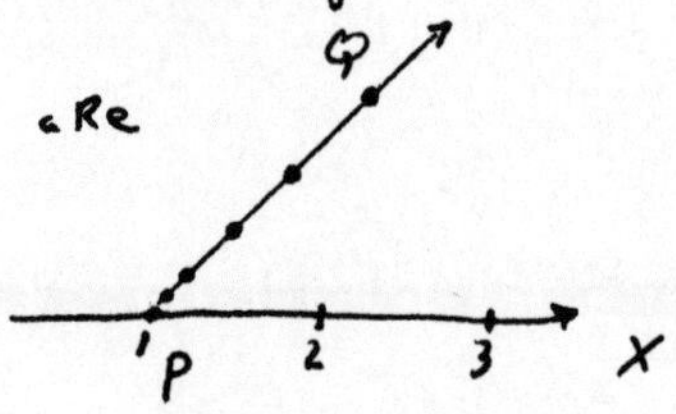

The Left-handed student would probably select the secant lines as follows:

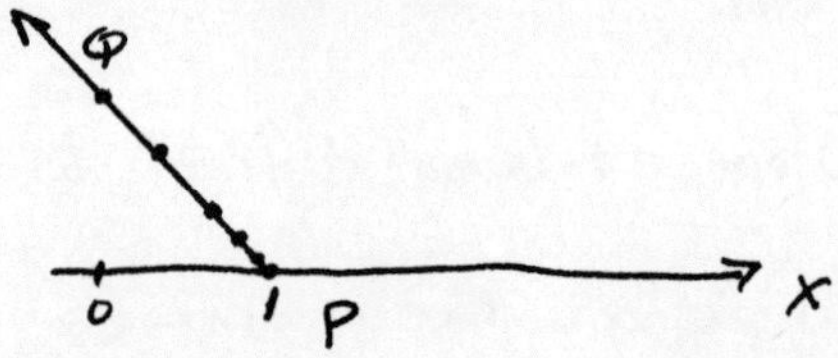

Clearly this student would conclude that the tangent line is $y = 1-x$. Since the tangent line must be unique there is __no__ __tangent__ at the point P (1,0). NOTE: A common ERROR is to think that the tangent is the x axis.

THE CHAIN RULE

3-17

Find h'(x) if

$$h(x) = \left(\frac{x + 1}{x - 2}\right)^4$$

****

Using the chain rule,

$$h'(x) = 4\left(\frac{x+1}{x-2}\right)^3 \cdot D_x\left(\frac{x+1}{x-2}\right)$$

$$= 4\left(\frac{x+1}{x-2}\right)^3 \cdot \left[\frac{(x-2)(1) - (x+1)(1)}{(x-2)^2}\right]$$

$$= 4\left(\frac{x+1}{x-2}\right)^3 \left(\frac{-3}{(x-2)^2}\right)$$

$$= -\frac{12(x+1)^3}{(x-2)^5}$$

3-18

Find $\frac{dy}{dx}$ if $y = \sqrt{1 + t^2}$ and $t = 3x + 1$.

****

$$\frac{dy}{dx} = \frac{dy}{dt} \cdot \frac{dt}{dx}$$

$$= \frac{1}{2}(1+t^2)^{-\frac{1}{2}} \cdot 2t \cdot \frac{d}{dx}(3x+1)$$

$$= \frac{3t}{\sqrt{1+t^2}}.$$

3-19

Given the function $f(x) = (x^2 - 4x + 4)^{1/3}$,
Evaluate $f'(x)$ and simplify your answer.

****

Method 1

$$f(x) = (x^2 - 4x + 4)^{1/3}$$

let $u = x^2 - 4x + 4$

$$\frac{du}{dx} = 2x - 4$$

$$f'(x) = \frac{1}{3}(x^2 - 4x + 4)^{-2/3}(2x - 4)$$

$$= \frac{1}{3}(x-2)^{-4/3}(2)(x-2)$$

$$= \boxed{2/3\,(x-2)^{-1/3}}$$

Method 2

$$f(x) = (x^2 - 4x + 4)^{1/3}$$

$$= (x-2)^{2/3}$$

$$\boxed{f'(x) = 2/3\,(x-2)^{-1/3}}$$

3-20

Differentiate $y = ((x^2 + 1)^2 + 1)^2 + 1$

****

$$\frac{dy}{dx} = 2((x^2+1)^2+1)(2)(x^2+1)(2x)$$

$$= 8x(x^2+1)(x^4+2x^2+2)$$

3-21

Use the chain rule to find dy/dx for $y = \sin^2(3x - 1)^2$.

****

$u = 3x - 1$
$v = u^2$
$w = \sin v$
$y = w^2$

$$\frac{dy}{dx} = \frac{dy}{dw} \cdot \frac{dw}{dv} \cdot \frac{dv}{du} \cdot \frac{du}{dx}$$

$$= (2w)(\cos v)(2u)(3)$$

$$= 12(\sin v)(\cos v)\,u$$
$$= 12(\sin u^2)(\cos u^2)\,u$$
$$= 12(\sin(3x-1)^2)(\cos(3x-1)^2)(3x-1)$$
$$= 12(3x-1)\sin(3x-1)^2\cos(3x-1)^2$$

3-22

Given

$$y = \frac{(x^2 + 3)^3}{(x^3 + 1)^4},$$

find the derivative of y with respect to x.

****

$$\frac{dy}{dx} = \frac{3(x^2+3)^2 2x(x^3+1)^4 - 4(x^3+1)^3 3x^2(x^2+3)^3}{(x^3+1)^8}$$

$$= \frac{6x(x^2+3)^2(x^3+1) - 12x^2(x^2+3)^3}{(x^3+1)^5}$$

$$= \frac{6x(x^2+3)^2[(x^3+1) - 2x(x^2+3)]}{(x^3+1)^5}$$

$$= \frac{6x(x^2+3)^2(x^3+1-2x^3-6x)}{(x^3+1)^5}$$

$$= -\frac{6x(x^2+3)^2(x^3+6x-1)}{(x^3+1)^5}$$

3-23

Differentiate (a) $\cos^3 x \cos 3x$ (b) $\sin(\cos\sqrt{x})$

(c) $\sec x \tan x$

****

(a) $\frac{d}{dx}\{\cos^3 x \cos 3x\} = (3\cos^2 x \cdot -\sin x)\cos 3x + (-\sin 3x \cdot 3)\cos^3 x$

$= -3\cos^2 x[\sin x \cos 3x + \sin 3x \cos x]$

$= -3\cos^2 x \sin(x+3x)$

$= \underline{-3\cos^2 x \sin 4x}$

(b) $\frac{d}{dx}\{\sin(\cos\sqrt{x})\} = \cos(\cos\sqrt{x}) \cdot -\sin\sqrt{x} \cdot \frac{1}{2\sqrt{x}}$

$= \underline{\frac{-1}{2\sqrt{x}}(\sin\sqrt{x})(\cos(\cos\sqrt{x}))}$

(c) $\frac{d}{dx}\{\sec x \tan x\} = (\sec x \tan x)\tan x + (\sec^2 x)\sec x$

$= \sec x(\tan^2 x + \sec^2 x)$

$\underline{\text{or } \sec x(1 + 2\tan^2 x)}$ using $\sec^2 x + 1 = \tan^2 x$

3-24

Calculate the derivative:

$$y = (4x^5 - 10)^{\frac{1}{6}}$$

****

Use power rule and the chain rule:

$$\frac{dy}{dx} = \frac{1}{6}(4x^5 - 10)^{-\frac{5}{6}}(20x^4) = \frac{10x^4}{3(4x^5 - 10)^{5/6}}$$

3-25

Find the derivative of $y = (x + 3)^{1/3}(2x - 1)^{2/3}$

****

$$\frac{dy}{dx} = (x+3)^{1/3} \cdot \frac{2}{3}(2x-1)^{-1/3} \cdot 2 + (2x-1)^{2/3} \cdot \frac{1}{3}(x+3)^{-2/3}$$

$$= \frac{4(x+3)^{1/3}}{3(2x-1)^{1/3}} + \frac{(2x-1)^{2/3}}{3(x+3)^{2/3}} = \frac{4(x+3) + (2x-1)}{3(2x-1)^{1/3}(x+3)^{2/3}}$$

$$= \frac{6x+11}{3(2x-1)^{1/3}(x+3)^{2/3}}$$

3-26

Differentiate each of the following:

a) $y = \sqrt{100-x^2}$

b) $y = \sqrt{x+\sqrt{100-x^2}}$

c) $y = \sqrt[3]{100-x^2}$

****

a) Substitute $u = 100 - X^2$ THEN $u' = -2X$ and

$$y' = \frac{u'}{2y} = \frac{-2X}{2y} = \frac{-X}{y} = \frac{-X}{\sqrt{100 - X^2}}.$$

b) Sub. $u = x$, $v = \sqrt{100 - X^2}$ and $\mathcal{U} = u + v$

THEN $u' = 1$, $v' = \dfrac{-X}{\sqrt{100 - X^2}}$ and $\mathcal{U}' = u' + v'$

(v' is by part a.)

ALSO $y' = \dfrac{\mathcal{U}'}{2y} = \dfrac{1 - \dfrac{X}{\sqrt{100 - X^2}}}{2y} = \dfrac{\dfrac{\sqrt{100 - X^2} - X}{\sqrt{100 - X^2}}}{2y}$

HENCE

$$y' = \frac{\sqrt{100 - X^2} - X}{2y\sqrt{100 - X^2}} = \frac{\sqrt{100 - X^2} - X}{2\sqrt{X + \sqrt{100 - X^2}}\sqrt{100 - X^2}}.$$

c) Sub. $u = 100 - X^2$ THEN $y = \sqrt[3]{u}$ and

$$y' = \frac{u'}{3y^2} = \frac{-2X}{3y^2} = \frac{-2X}{3(100 - X^2)^{2/3}}$$

THE DERIVATIVE AS A RATE OF CHANGE

3-27

If distance, s, in feet, of an object from a starting point, after t seconds is given by $s = 5t - 0.4t^2$, find (1) the object's velocity after three seconds, (2) the time at which the object will stop, and (3) the object's acceleration after five seconds.

****

(1) $v = \frac{ds}{dt} = 5 - 0.8t$

AT $t = 3$, $v = 5 - 0.8(3) = 2.6$ ft/sec

(2) $5 - 0.8t = 0$

$t = 6.25$ sec.

(3) $a = \frac{d^2s}{dt^2} = -0.8$ ft/sec²

3-28

Suppose that $y = 2x^3 - 3x$.

(a) Find the instantaneous rate at which y changes with x when $x = 2$.

(b) Find the average rate at which y changes with x between $x = 1$ and $x = 3$.

****

(a) $\frac{dy}{dx} = 6x^2 - 3$ and the answer is $\left.\frac{dy}{dx}\right|_{x=2} = 21$.

(b) answer is $\frac{\Delta y}{\Delta x} = \frac{(y \text{ when } x=3) - (y \text{ when } x=1)}{3 - 1}$

$$= \frac{45 - (-1)}{2} = 23.$$

3-29

Acme Company has studied the productivity of its employees and found that after t years of experience an employee can produce

$$p(t) = -t^2 + 60t + 100 \text{ units per year.}$$

Find the rate of change of employee productivity and discuss the employee productivity of an employee with 10 years of experience, with 30 years of experience.

****

The rate of change of employee productivity is given by the derivative

$$p'(t) = -2t + 60.$$

When $t = 10$, $p'(10) = 40$, so an employee with exactly 10 years of experience will be increasing his or her productivity at a rate of 40 units per year.

When $t = 30$, $p'(30) = 0$, so an employee with exactly 30 years of experience will find his or her productivity neither increasing nor decreasing.

3-30

An arrow is shot straight up into the air at time t= 0 (seconds). Its displacement at time t is $S(t)= 100-32t^2$ feet. What is the velocity at time t?

****

By the definition of velocity, the velocity of the arrow is the derivative of S(t) with respect to t. So the velocity, V(t) is given by the following:

$$V(t) = \frac{ds}{dt}$$
$$= -64t \text{ feet per sec}$$

3-31

A spherical balloon is being inflated in such a way that its radius is increasing at the constant rate of 3 cm/min. If the volume of the balloon is 0 at time 0, at what rate is the volume increasing after 5 minutes?

****

LET r BE THE LENGTH OF THE RADIUS. THEN r AT TIME t IS GIVEN BY

$$r = 3t.$$

THE VOLUME IS GIVEN BY

$$V = \frac{4}{3}\pi r^3 = \frac{4}{3}\pi(3t)^3 = 36\pi t^3$$

SO

$$\frac{dV}{dt} = 108\pi t^2$$

IS THE RATE OF CHANGE OF VOLUME.

AT TIME $t = 5$,

$$\frac{dV}{dt}(5) = 108\pi(5)^2 = 2700\pi.$$

THE VOLUME IS INCREASING AT 2700π CM^3/MIN.

3-32

The volume in liters of antifreeze mixture remaining in a leaking radiator after t minutes is $V = 20 - 2t - 0.1t^2$.

(a) How long does it take the antifreeze mixture to drain from the radiator?

(b) How fast is the volume changing when $t = 5$?

****

(a) SET $V = 0$ AND SOLVE FOR t.

$$-\frac{1}{10}t^2 - 2t + 20 = 0$$

$$t^2 + 20t - 200 = 0$$

$$t = \frac{-b \pm \sqrt{b^2 - 4ac}}{2a} = \frac{-20 \pm \sqrt{(20)^2 - 4(-200)}}{2} = \frac{-20 \pm \sqrt{1200}}{2}$$

$$= \frac{-20 \pm 20\sqrt{3}}{2} = -10 \pm 10\sqrt{3}$$

BUT $t > 0$. $\therefore$ $t = -10 + 10\sqrt{3}$ MINUTES

(b) $$\left.\frac{dV}{dt}\right|_{t=5} = \Big[-2 - 0.1(2t)\Big]_{t=5} = -2 - .2(5) = -2 - 1$$

$$= -3 \text{ liters/min.}$$

IMPLICIT DIFFERENTIATION

3-33

Find the derivative $\frac{dp}{dt}$ for $P = \frac{mv}{\sqrt{1 - \frac{v^2}{c^2}}}$, where m and c are constants.

****

$$p = \frac{mv}{\sqrt{1-\frac{v^2}{c^2}}} = mv\left(1-\frac{v^2}{c^2}\right)^{-\frac{1}{2}}$$

$$\frac{dP}{dt} = mv\left(-\frac{1}{2}\right)\left(1-\frac{v^2}{c^2}\right)^{-\frac{3}{2}}\left(-\frac{2v}{c^2}\frac{dv}{dt}\right) + m\frac{dv}{dt}\left(1-\frac{v^2}{c^2}\right)^{-\frac{1}{2}}$$

$$= \frac{m\frac{v^2}{c^2}\frac{dv}{dt}}{\left(1-\frac{v^2}{c^2}\right)^{\frac{3}{2}}} + \frac{m\frac{dv}{dt}}{\left(1-\frac{v^2}{c^2}\right)^{\frac{1}{2}}}$$

$$= m\frac{dv}{dt}\left(\frac{\frac{v^2}{c^2}}{\left(1-\frac{v^2}{c^2}\right)^{\frac{3}{2}}} + \frac{1}{\left(1-\frac{v^2}{c^2}\right)^{\frac{1}{2}}}\cdot\frac{\left(1-\frac{v^2}{c^2}\right)}{\left(1-\frac{v^2}{c^2}\right)}\right)$$

$$= m\frac{dv}{dt}\left(\frac{\frac{v^2}{c^2} + 1 - \frac{v^2}{c^2}}{\left(1-\frac{v^2}{c^2}\right)^{\frac{3}{2}}}\right)$$

$$= m\frac{dv}{dt}\frac{1}{\left(1-\frac{v^2}{c^2}\right)^{\frac{3}{2}}}$$

$$\frac{dP}{dt} = \frac{m}{\left(1-\frac{v^2}{c^2}\right)^{\frac{3}{2}}}\frac{dv}{dt}.$$

3-34

Find the equation of the line normal to the curve defined by the equation:

$$x^3y^4 - 5 = x^3 - x^2 + y,$$

at the point (2,-1).

****

Differentiate implicitly:

$$3x^2y^4 + 4x^3y^3\, dy/dx = 3x^2 - 2x + dy/dx$$

$$dy/dx\,(4x^3y^3 - 1) = 3x^2 - 2x - 3x^2y^4$$

$$dy/dx = \left.\frac{3x^2 - 2x - 3x^2y^4}{4x^3y^3 - 1}\right|_{\substack{x=2\\y=-1}} = \frac{12-4-12}{-32-1}$$

$$= 4/33$$

∴ Normal line at (2,−1) has slope −33/4.

Equation of normal line: $y+1 = \frac{-33}{4}(x-2)$

or: $33x + 4y - 62 = 0$

3-35

Use implicit differentiation to find y" if $2xy = y^2$. Simplify your answer and leave it in terms of x and y.

****

$$2xy = y^2$$

$$(2x)y' + y(2) = 2yy'$$

solve for y': $xy' + y = yy'$

$$xy' - yy' = -y$$

$$y'(x-y) = -y$$

$$y' = \frac{-y}{x-y} = \frac{y}{y-x}$$

find y'' implicitly

$$y'' = \frac{(y-x)y' - y(y'-1)}{(y-x)^2}$$

simplify: $y'' = \frac{yy' - xy' - yy' + y}{(y-x)^2} = \frac{y - xy'}{(y-x)^2}$

substitute for y':

$$y' = \frac{y - x\left(\frac{y}{y-x}\right)}{(y-x)^2} = \frac{y(y-x) - xy}{(y-x)^3}$$

$$= \frac{y^2 - xy - xy}{(y-x)^3} = \frac{y^2 - 2xy}{(y-x)^3}$$

3-36

Find $D_x y$ if $3y^2 + x^2y^3 + x^3 = 0$.

****

$$d(3y^2 + x^2y^3 + x^3) = d(0)$$

$$d(3y^2) + d(x^2y^3) + d(x^3) = 0$$

$$6y\,dy + [x^2 \cdot 3y^2\,dy + y^3 \cdot 2x\,dx] + 3x^2\,dx = 0$$

$$6y\,dy + 3x^2y^2\,dy + 2xy^3\,dx + 3x^2\,dx = 0$$

$$(6y + 3x^2y^2)\,dy = -(3x^2 + 2xy^3)\,dx$$

$$\frac{dy}{dx} = -\frac{(3x^2 + 2xy^3)}{(6y + 3x^2y^2)}$$

$$\therefore\ D_x y = \frac{dy}{dx} = -\frac{x(3x + 2y^3)}{3y(2 + x^2y)}.$$

3-37

Given

$$x^3 + 3x^2y + y^3 = 4,$$

use implicit differentiation to find the derivative of y with respect to x.

****

$$3x^2 + 6xy + 3x^2\frac{dy}{dx} + 3y^2\frac{dy}{dx} = 0$$

$$(3x^2 + 3y^2)\frac{dy}{dx} = -3x^2 - 6xy$$

$$\frac{dy}{dx} = \frac{-3x^2 - 6xy}{3x^2 + 3y^2}$$

$$\frac{dy}{dx} = -\frac{x^2 + 2xy}{x^2 + y^2}$$

3-38

Find D_xy if $3x^2 + xy + 5y^2 = 4$.

****

Differentiating each term with respect to x, we get

$$D_x(3x^2) + D_x(xy) + D_x(5y^2) = 0;$$

by the use of the power rule and the product rule, we obtain

$$6x + x(D_xy) + y + 10y(D_xy) = 0.$$

Then

$$D_xy\,(x + 10y) = -6x - y$$

and

$$D_xy = \frac{-6x - y}{x + 10y}$$

3-39

Use implicit differentiation to find $\frac{dy}{dx}$ for the curve given by $2y^3 - 3xy^2 + 1 = x^2 - y$.

****

$$6y^2y' - 3(y^2 + 2xyy') + 0 = 2x - y'$$

Now solve for y':

$$6y^2y' - 3y^2 - 6xyy' = 2x - y'$$

$$6y^2y' - 6xyy' + y' = 2x + 3y^2$$

$$(6y^2 - 6xy + 1)\,y' = 2x + 3y^2$$

So, $$y' = \frac{2x + 3y^2}{6y^2 - 6xy + 1}$$

3-40

Find $\frac{dy}{dx}$ if $x + y^2 = xy^2$

****

$$1 + 2y\frac{dy}{dx} = x\left(2y\frac{dy}{dx}\right) + y^2$$

$$\frac{dy}{dx} = \frac{y^2 - 1}{2y - 2xy}$$

3-41

a) Sketch the graphs of $x^2 + 2y^2 = 6$ and $x^2 = 4y$ on the same coordinate system and find their points of intersection.

b) Use implicit differentiation to show that the curves in part (a) are orthogonal.

[Note: Two curves are said to be orthogonal if their tangent lines at the point of intersection are perpendicular.]

****

a)

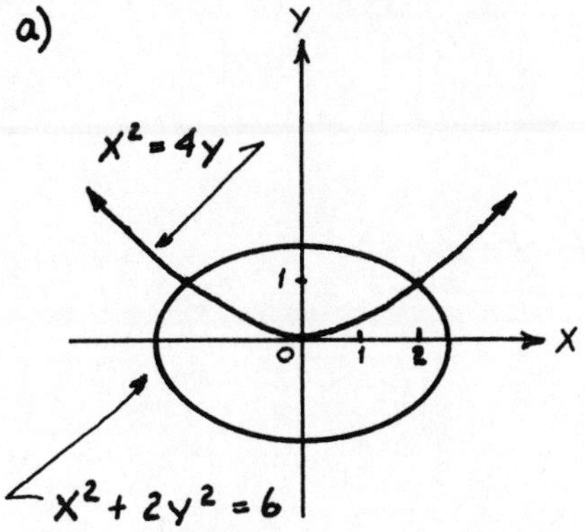

To find the points of intersection substitute $x^2 = 4y$ into $x^2 + 2y^2 = 6$ giving:

$$4y + 2y^2 = 6 \Rightarrow y^2 + 2y - 3 = 0$$
$$\Rightarrow (y+3)(y-1) = 0 \Rightarrow y = -3 \text{ or } 1$$

For $y = -3$, $x^2 = -12$ (no solution)
and $y = 1$, $x^2 = 4 \Rightarrow x = \pm 2$

The curves intersect at $\underline{\underline{(2,1) \text{ and } (-2,1).}}$

b) For $x^2 + 2y^2 = 6$, $2x + 4y\dfrac{dy}{dx} = 0$, so $\dfrac{dy}{dx} = \dfrac{-x}{2y}$.

For $x^2 = 4y$, $2x = 4\dfrac{dy}{dx}$, so $\dfrac{dy}{dx} = \dfrac{x}{2}$

At $(2,1)$ the slopes of the tangents are -1 and 1 so they are perpendicular at this point.

At $(-2,1)$ the slopes are 1 and -1 showing again that the lines are perpendicular.

Hence the curves are orthogonal at $(2,1)$ and $(2,-1)$.

3-42

Find y" if $xy + y^3 = 1$.

****

$y + xy' + 3y^2y' = 0$, so

$$y' = \frac{-y}{x+3y^2}.$$

$$y'' = \frac{-y'(x+3y^2) + y(1+6yy')}{(x+3y^2)^2}$$

$$= \frac{2y + 6y^2y'}{(x+3y^2)^2}$$

$$= \frac{2y - 6y^2 \cdot \frac{y}{x+3y^2}}{(x+3y^2)^2}$$

$$= \frac{2xy + 6y^3 - 6y^3}{(x+3y^2)^3}$$

$$= \frac{2xy}{(x+3y^2)^3}.$$

3-43

What is the slope of the tangent line to the curve $xy^3 + y - 1 = 0$ at the point (0,1)?

****

By substitution, note that the point (0,1) does indeed lie on the given curve, so the problem makes sense.

The slope of the tangent line to the curve at (0,1) is given by: $\left.\frac{dy}{dx}\right]_{\substack{x=0\\y=1}}$

Differentiating both sides of the equation

$xy^3 + y - 1 = 0$ with respect to x, gives:

$$x \cdot 3y^2 \cdot \frac{dy}{dx} + y^3 + \frac{dy}{dx} = 0$$

Substituting $x=0$, $y=1$ above, gives

$$0 + 1 + \left.\frac{dy}{dx}\right]_{\substack{x=0\\y=1}} = 0, \text{ so } \left.\frac{dy}{dx}\right]_{\substack{x=0\\y=1}} = -1$$

The slope of the tangent to the curve at the point (0,1) is: -1

3-44

Find the derivative dy/dx (where it exists) for the following:

$$2x + xy = y^4$$

****

Implicit Differentiation : $\frac{d}{dx}(2x + xy) = \frac{d}{dx}(y^4)$

$$\Rightarrow \quad \frac{d}{dx}(2x) + \frac{d}{dx}(xy) = \frac{d}{dx}(y^4)$$

$$\frac{d}{dx}(2x) = 2, \quad \frac{d}{dx}(xy) = \underbrace{x\frac{dy}{dx} + y\frac{dx}{dx}}_{\text{product rule}} = x\frac{dy}{dx} + y,$$

$$\frac{d}{dx}(y^4) = 4y^3\frac{dy}{dx} \quad \text{(Chain rule)}$$

$$\Rightarrow \quad 2 + x\frac{dy}{dx} + y = 4y^3\frac{dy}{dx} \Rightarrow 2 + y = (4y^3 - x)\frac{dy}{dx}$$

$$\Rightarrow \quad \frac{dy}{dx} = \frac{2 + y}{4y^3 - x}$$

3-45

Find $\frac{dy}{dx}$ for $x^2 + xy^2 = 6$ by differentiating implicitly; then calculate the value of this derivative when $x = 2$ and $y > 0$.

****

Differentiating implicitly:

$$2x + x \cdot 2y\frac{dy}{dx} + y^2 = 0$$

When $x = 2$

$$2^2 + 2y^2 = 6$$

$$y = 1 \text{ or } -1$$

Since $y > 0$ $y = 1$

To find $\frac{dy}{dx}$ when $x = 2$ and $y = 1$:

$$(2)(2) + 2 \cdot 2(1)\frac{dy}{dx} + 1^2 = 0$$

$$\frac{dy}{dx} = -\frac{5}{4}$$

3-46

If c is a constant and $p^2q - 3pq^2 = 4213c^2 - 333$, find $\frac{dp}{dq}$.

****

Note since c is a constant, $4213c^2 - 333$ is also a constant. Differentiating implicitly with respect to q,

$$2pq\frac{dp}{dq} + p^2 - 3p \cdot 2q - 3q^2\frac{dp}{dq} = 0.$$

So $(2pq - 3q^2)\frac{dp}{dq} = 6pq - p^2$ or $\frac{dp}{dq} = \frac{6pq - p^2}{2pq - 3q^2}$.

3-47

Find $\frac{dy}{dx}$ if sin (2x + 3y) = 3xy + 5y -2

****

$$\cos(2x+3y)\left[2+3\frac{dy}{dx}\right] = 3x\frac{dy}{dx} + 3y + 5\frac{dy}{dx}$$

$$2\cos(2x+3y) + 3\frac{dy}{dx}\left(\cos(2x+3y)\right) = 3x\frac{dy}{dx} + 3y + 5\frac{dy}{dx}$$

$$3\frac{dy}{dx}\cos(2x+3y) - 3x\frac{dy}{dx} - 5\frac{dy}{dx} = 3y - 2\cos(2x+3y)$$

$$\Rightarrow \frac{dy}{dx} = \frac{3y - 2\cos(2x+3y)}{3\cos(2x+3y) - 3x - 5}$$

3-48

Given $x^2 + y^2 - 2x + 4y = -9$ a) find the derivative y'by proceeding in the usual way and b) show that your result is nonsense.
Hint: The assumption that y represents an implicit function of x is not valid.

****

a) $2x + 2yy' - 2 + 4y' = 0$ Add $2-2x$ Then $(2y+4)y' = 2-2x$ and $y' = \frac{2-2x}{2y+4}$ or $\frac{1-x}{y+2}$

b) Given $x^2 + y^2 - 2x + 4y = -9$. Rewrite as

$x^2 - 2x + y^2 + 4y = -9$. Complete the square:

$x^2 - 2x + \underline{1} + y^2 + 4y + \underline{4} = -9 + \underline{5}$ Then

$(x-1)^2 + (y+2)^2 = -4$ has no Graph, since the left side is positive and the right side is negative. (y cannot represent a function of x even implicitly!)

THE TRIGONOMETRIC FUNCTIONS

3-49

Find y' if $y = 3 \sin 2x \cos^2 4x$.

****

$$y' = 6\cos 2x \cos^2 4x - 6 \sin 2x \cos 4x \sin 4x \cdot 4$$

$$= 6 \cos 4x (\cos 2x \cos 4x - 4 \sin 2x \sin 4x)$$

3-50

For each of the two functions given below, find dy/dx. Identify the difference in form between the two functions.

function A: $y = (\sin x)(\cos x)$ function B: $y = \sin(\cos x)$

****

In A, the function is a product of two trig functions. To find dy/dx use the product rule.

$$\frac{dy}{dx} = (\sin x)(-\sin x) + (\cos x)(\cos x)$$
$$= -\sin^2 x + \cos^2 x$$
$$= \cos^2 x - \sin^2 x = \cos 2x$$

In B, the function is a composite of two trig functions. To find dy/dx use the chain rule.

Let $u = \cos x$ then $y = \sin u$

$$\frac{dy}{dx} = \frac{dy}{du} \cdot \frac{du}{dx} = (\cos u)(-\sin x)$$
$$= -(\cos(\cos x))(\sin x)$$

3-51

Explain why the following statement is false.

If $f(u) = \sec(u)$, then $f'(u) = \sec(u)\tan(u)\dfrac{du}{dx}$.

****

The notation $f'(u)$ means the derivative of $f(u)$ with respect to u.

To make the statement true

a) the derivative wrt $x \to f'(u)\dfrac{du}{dx} = \sec u \tan u \dfrac{du}{dx}$

b) the derivative wrt $u \to f'(u) = \sec u \tan u$

3-52

Find the first derivative of the function $f(x) = \sqrt{\sin x}$.

****

$$f(x) = (\sin x)^{\frac{1}{2}}$$

$$\therefore f'(x) = \frac{1}{2}(\sin x)^{-\frac{1}{2}}\cos x = \frac{\cos x}{2\sqrt{\sin x}}.$$

3-53

Find $f'(x)$ if $f(x) = x\cos(\frac{1}{x})$.

****

Use the product rule:

$$f'(x) = x\cdot\left(-\sin\left(\tfrac{1}{x}\right)\right)\cdot\frac{-1}{x^2} + \cos\left(\tfrac{1}{x}\right) = \cos\left(\tfrac{1}{x}\right) + \frac{1}{x}\sin\left(\tfrac{1}{x}\right).$$

3-54

Let x = sin(y). Find $\frac{dy}{dx}$ in terms of x.

****

If $x = \sin y$, differentiating implicitly with respect to x we get

$$1 = \cos y \cdot dy/dx$$

$$\Rightarrow dy/dx = \frac{1}{\cos y} = \frac{\pm 1}{\sqrt{1-\sin^2 y}}$$

$$= \pm \frac{1}{\sqrt{1-x^2}}$$

3-55

Find the slope of $f(t) = \sqrt{1-\sin 2t}$ at (a) the point (0,1). (b) the origin.

****

First find the derivative $f'(t)$:

SUB. $y = f(t)$ and $u = 1-\sin 2t$ THEN

$$y = f(t) = \sqrt{u} \quad \text{and} \quad y' = \frac{u'}{2y} = \frac{-2\cos 2t}{2y}$$

$$\text{OR} \quad f'(t) = \frac{-\cos 2t}{\sqrt{1-\sin 2t}}$$

(a) $f'(0) = \frac{-\cos 0}{\sqrt{1-\sin 0}} = \frac{-1}{\sqrt{1}} = -1$

(b) The Origin is not on the graph!

THE DERIVATIVE OF THE POWER FUNCTION FOR RATIONAL EXPONENTS

3-56

A line is tangent to the curve $y = \dfrac{x^2 + 3}{(x+3)^{1/2}}$ at the point where $x = 1$. Write the equation of this line.

****

To write equation of a line, we need point and slope.

To get point: If $x = 1$ $\quad y = \dfrac{1^2+3}{(1+3)^{1/2}} = 2$ $\quad$ point is $(1,2)$

To get slope: $\dfrac{dy}{dx} = \dfrac{(x+3)^{\frac{1}{2}}\cdot 2x - (x^2+3)\frac{1}{2}(x+3)^{-\frac{1}{2}}}{x+3}$

$$\left.\frac{dy}{dx}\right|_{x=1} = \frac{(1+3)^{\frac{1}{2}}\cdot 2\cdot 1 - (1^2+3)\frac{1}{2}(1+3)^{-\frac{1}{2}}}{1+3} = \frac{3}{4}$$

Equation of line is: $y - 2 = \frac{3}{4}(x-1)$

or $\quad y = \frac{3}{4}x + \frac{5}{4}$

3-57

Find y' if $y = (2x^3 + 3x)^{3/2}$.

****

$$y' = \frac{3}{2}(2x^3+3x)^{\frac{3}{2}-1}\cdot(6x^2+3)$$

$$= \left(9x^2 + \frac{9}{2}\right)\sqrt{2x^3+3x}\,.$$

3-58

Find an equation of the tangent line to $y = \sqrt{25 - x^2}$ at the point (3,4).

****

The slope of the tangent is the derivative evaluated at the point. Write

$$y = (25 - x^2)^{1/2}$$

and differentiate using the chain rule.

$$y' = \frac{1}{2}(25 - x^2)^{1/2 - 1} \frac{d}{dx}(25 - x^2)$$

$$= \frac{1}{2}(25 - x^2)^{-1/2}(-2x) = \frac{-x}{(25 - x^2)^{1/2}}$$

At the point (3,4),

$$y' = \frac{-3}{(25-9)^{1/2}} = \frac{-3}{\sqrt{16}} = -\frac{3}{4}$$

Using the point-slope form of a line, the equation is

$$-\frac{3}{4} = \frac{y-4}{x-3}, \quad -3x + 9 = 4y - 16,$$

$$3x + 4y = 25$$

You may want to write the equation in slope-intercept form.

$$y = -\frac{3}{4}x + \frac{25}{4}$$

DERIVATIVES OF HIGHER ORDER

3-59

Explain why the following statement is false.

Notations for the fourth derivative are $\frac{d^4y}{dx^4}$ and $f^4(x)$.

****

$\frac{d^4y}{dx^4}$ is acceptable notation for $\frac{d^4}{dx^4}(y)$.

$f^4(x)$ means $[f(x)]^4$ not $f''''(x) = f^{(4)}(x)$

3-60

Let $f(x) = x/(x - 1)$. Find $f'(x)$, $f''(x)$, and a formula for $f^{(n)}(x)$.

****

Use the quotient rule for $f'(x)$.

$$f'(x) = \frac{(x-1)\frac{d}{dx}x - x\frac{d}{dx}(x-1)}{(x-1)^2} = \frac{(x-1)(1) - x(1)}{(x-1)^2}$$

$$= \frac{-1}{(x-1)^2} = -(x-1)^{-2}$$

The negative exponent form is convenient, because it allows us to use the chain rule to find $f''(x)$.

$$f''(x) = -(-2)(x-1)^{-3} = 2(x-1)^{-3}$$

Although the instructions do not require it, let us find a few more higher derivatives to establish a pattern.

$$f^{(3)}(x) = 2(-3)(x-1)^{-4} = -(3)(2)(x-1)^{-4}$$

$$f^{(4)}(x) = -(3)(2)(-4)(x-1)^{-5} = (4)(3)(2)(x-1)^{-5}$$

A pattern should now be clear. $f^{(n)}(x)$ will be of form

$$\pm n(n-1)(n-2)\cdots(2)(x-1)^{-(n+1)}$$

Recognize $n(n-1)(n-2)\cdots(2)$ as $n!$. Odd numbered derivatives will be negative, even numbered ones positive. A factor of $(-1)^n$ will cause the correct sign. Thus

$$f^{(n)}(x) = (-1)^n n!\,(x-1)^{-(n+1)}$$

This result could be proven for any positive integer n using mathematical induction.

DIFFERENTIABILITY AND CONTINUITY

3-61

Given that f'(x) is continuous at x=a, show that f(x) is differentiable at x=a.

****

Recall that $f(x)$ continuous at $x=a$ means that

$\lim_{x\to a} f(x) = f(a)$ and $f(a)$ exists. The Proof follows:

We are given that $f'(x)$ is continuous at $x=a$. If we substitute $\mathcal{F}(x) = f'(x)$ this means that $\lim\limits_{x \to a} \mathcal{F}(x) = \mathcal{F}(a)$ and $\mathcal{F}(a)$ exists. However by back substitution this means that $f'(a)$ exists.

CAUTION! $f(x)$ continuous at $x=a$ does not imply that $f(x)$ is differentiable at $x=a$.

■■ **3-62**

Determine a and b so the following function is everywhere differentiable.

$$f(x) = \begin{cases} x^2 + 3x & \text{if } x \geq 2 \\ ax + b & \text{if } x < 2 \end{cases}$$

****

Differentiable: $\lim\limits_{x \to 2^+} f'(x) = \lim\limits_{x \to 2^-} f'(x)$

$\left. \begin{array}{l} \lim\limits_{x \to 2^+} f'(x) = \lim\limits_{x \to 2^+} (2x+3) = 7 \\[2mm] \lim\limits_{x \to 2^-} f'(x) = \lim\limits_{x \to 2^-} a = a \end{array} \right\} \quad a = 7$

Continuous: $\lim\limits_{x \to 2^+} f(x) = \lim\limits_{x \to 2^-} f(x) = f(2)$

$\left. \begin{array}{l} \lim\limits_{x \to 2^+} f(x) = \lim\limits_{x \to 2^+} (x^2+3x) = 10 \\[2mm] \lim\limits_{x \to 2^-} f(x) = \lim\limits_{x \to 2^-} (7x+b) = 14+b \\[2mm] f(2) = 2^2 + 3(2) = 10 \end{array} \right\} \quad b = -4$

3-63

a) Sketch a graph of a function which is continuous on (2,4) such that $f(2) = f(4) = 2$; $f(3) = 4$; $f'(3) = 0$.

b) Sketch a graph of a function which is continuous on (1,5) such that $f(2) = f(4) = 2$; $f(3) = 4$; but $f'(x) \neq 0$ for any x in (2,4).

****

Note the solutions are not unique

a)

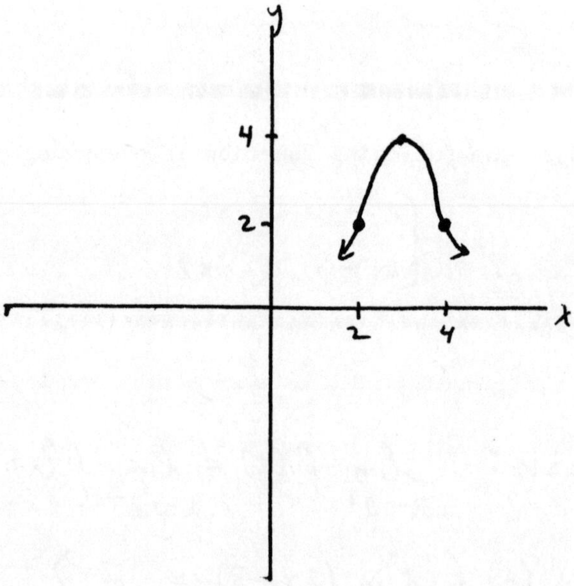

b)

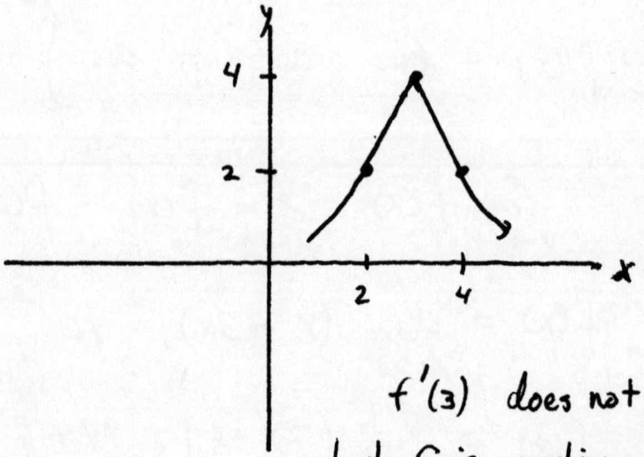

$f'(3)$ does not exist

but f is continuous at $x=3$

3-64

Define a function g(x) by: $g(x) = \begin{cases} x^2 \sin \frac{1}{x}, & x \neq 0 \\ 0, & x = 0 \end{cases}$

Find g'(0) if it exists. Is g'(x) continuous for all values of x?

****

$$g'(0) = \lim_{\Delta x \to 0} \frac{g(0+\Delta x) - g(0)}{\Delta x} = \lim_{\Delta x \to 0} \frac{(\Delta x)^2 \sin \frac{1}{\Delta x} - 0}{\Delta x}$$

$$= \lim_{\Delta x \to 0} \Delta x \sin \frac{1}{\Delta x} = 0$$

$\therefore$ g'(0) exists, and $g'(0) = 0$.

For $x \neq 0$, $g'(x) = 2x \sin \frac{1}{x} + x^2(-x^{-2}) \cos \frac{1}{x}$

$$= 2x \sin \frac{1}{x} - \cos \frac{1}{x}$$

Clearly, $\lim_{x \to 0} g'(x) = \lim_{x \to 0} \left[2x \sin \frac{1}{x} - \cos \frac{1}{x} \right]$

which does not exist, hence, g'(x) is not continuous at $x = 0$. However, g'(x) is continuous for all other values of x.

3-65

Let $h(x) = \begin{cases} x + 4, & x \le -2 \\ 5 - x^2, & -2 < x \le 1 \\ 2x + 2, & x > 1 \end{cases}$. Discuss all points of discontinuity for h(x).

****

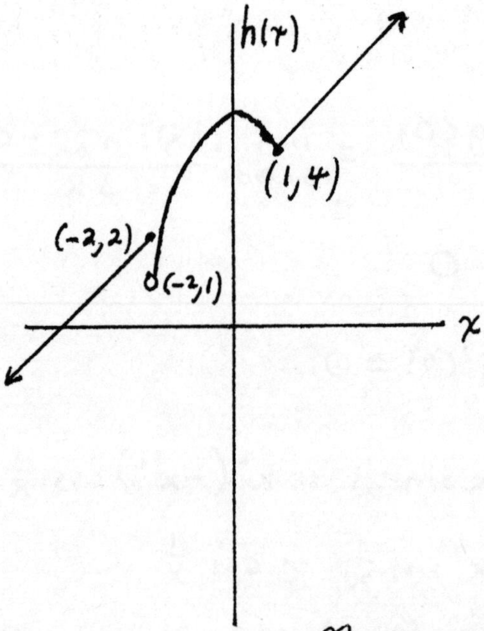

Note that $\lim_{x \to -2^-} h(x) = 2$, while $\lim_{x \to -2^+} h(x) = 1$. Hence there is a discontinuity at $x = -2$.

4
APPLICATIONS OF DIFFERENTIATION

DIFFERENTIALS AND APPROXIMATIONS

4-1

Approximate the fourth root of 16.6 using differentials.

****

Let $y = x^{1/4}$. $dy = \frac{1}{4}x^{-3/4}\,dx$. When $x = 16$, $y = 2$.

When $dx = .6$, $dy = \frac{1}{4}(16)^{-3/4}(.6) = \frac{1}{4}\left(\frac{1}{8}\right)(.6) = .01875$

Therefore when $x = 16 + .6$, $y \approx 2 + .01875$, so

$\sqrt[4]{16.6} \approx 2.01875$

4-2

If the sides of a cube are measured with an error of 2%, use differentials to estimate the relative error in the volume.

****

Let ΔV be the true (absolute) error in the volume V

then $\dfrac{\Delta V}{V}$ = true (relative) error in V

$\dfrac{\Delta V}{V}$ is approximated by $\dfrac{dV}{V}$

$V = s^3$, so $dV = \dfrac{dV}{ds} \cdot ds = 3s^2 ds$

$$\therefore \frac{dV}{V} = \frac{3s^2}{V} ds = \frac{3s^2}{s^3} \cdot ds = 3 \cdot \frac{ds}{s}$$

Since $\dfrac{ds}{s} = \pm 2\%$, $\dfrac{dV}{V} = \pm 6\%$

4-3

Using differentials, approximate $\sqrt[3]{10}$.

****

LET $f(x) = x^{1/3}$,
THEN $f'(x) = \frac{1}{3} x^{-2/3}$.

WE ASSUME $f(x+\Delta x) \doteq f(x) + f'(x)\Delta x$
SO
$$\begin{aligned} f(8+2) &\doteq f(8) + f'(8)2 \\ &= 8^{1/3} + \tfrac{1}{3} \cdot 8^{-2/3} \cdot 2 \\ &= 2 + \tfrac{1}{3} \cdot \tfrac{1}{4} \cdot 2 \\ &= 2\tfrac{1}{6}. \end{aligned}$$

THEREFORE, $\sqrt[3]{10} \doteq 2\frac{1}{6}$.

4-4

Use differentials to get a fractional approximation for $\sqrt[3]{7.9}$.

****

We let $y = x^{\frac{1}{3}}$. We know that $\sqrt[3]{8} = 2$, so we will have $x = 8$, $y = 2$, and $dx = -0.1$.

$$y' = \frac{1}{3}x^{-\frac{2}{3}} = \frac{1}{3x^{2/3}}$$

Then $dy = y'dx = \frac{1}{3x^{\frac{2}{3}}}dx = \frac{1}{3\cdot 8^{2/3}}(-.1)$

$$= \frac{-.1}{3(4)} = -\frac{.1}{12} = -\frac{1}{120}$$

So $y + dy = 2 + \left(-\frac{1}{120}\right) = 1\frac{119}{120} \approx \sqrt[3]{7.9}$.

4-5

Approximate $\sqrt[4]{16.3}$ using derivatives (differentials).

****

Let $f(x) = \sqrt[4]{x} = x^{1/4}$.

Then $f(x + \Delta x) = f(x) + \Delta y \approx f(x) + dy = f(x) + f'(x)dx$.

If $x = 16$, $f(x) = 2$. Take $\Delta x = dx = .3$.

$f'(x) = \frac{1}{4}x^{-3/4}$, so $f'(16) = \frac{1}{4}\cdot\frac{1}{(\sqrt[4]{16})^3} = \frac{1}{4}\cdot\frac{1}{8} = \frac{1}{32}$.

Thus $f(16.3) \approx f(16) + f'(16)dx = 2 + \frac{1}{32}(.3) = 2\frac{3}{320}$.

4-6

Use differentials to approximate the change in f(x) if x changes from 3 to 3.01 and $f(x) = (3x^2 - 26)^{10}$.

****

$$dy\Big|_{\substack{x=3\\dx=.01}} = f'(x)\,dx\Big|_{\substack{x=3\\dx=.01}} = 10(3x^2-26)^9 6x\,dx\Big|_{\substack{x=3\\dx=.01}} = 10(1)^9(6)(3)(.01) = 1.8$$

4-7

Using differentials find an approximation for $\sqrt[4]{83}$.

****

SET-UP $y = f(x) = \sqrt[4]{x}$ WITH $x = 81$, $\Delta x = dx = 2$

(THINK OF $81 = 3^4$ AS A BASE POINT.)

By definition $\Delta y = f(x+\Delta x) - f(x)$ hence

$\Delta y = f(83) - f(81) = f(83) - 3$ and

$f(83) = \Delta y + 3$ OR $\sqrt[4]{83} = \Delta y + 3$

ASSUME $\Delta y \doteq dy$ THEN $\sqrt[4]{83} \doteq dy + 3$

($\doteq$ means approximately equal.)

By def. $dy = f'(x)\,dx = \dfrac{dx}{4x^{3/4}}$

TO COMPUTE dy SUB. $dx = \Delta x = 2$ and $x = 81$.

THEN $dy = \dfrac{2}{4(81)^{3/4}} = \dfrac{2}{4 \cdot 3^3} = \dfrac{1}{54}$ HENCE

$\sqrt[4]{83} \doteq 3\,{}^1\!/_{54}$ OR $\underline{\underline{3.0185}}$

(Use your calculator to check for accuracy)

4-8

Use a differential to estimate $34^{\frac{1}{2}}$.

****

Let $f(x) = x^{1/2}$, $x = 36$ and $dx = -2$

$$f(34) \approx f(36) + f'(36)dx = \sqrt{36} + \frac{1}{2\sqrt{36}}\cdot(-2)$$

$$34^{1/2} \approx 6 - \frac{1}{6} = \frac{35}{6} \approx 5.8$$

4-9

Given the function $f(x) = x^{-1/3}$,

(a) find df, the differential of f, and

(b) approximate $\frac{1}{\sqrt[3]{7.952}}$ using the result in part (a).

****

(a) $df = f'(x)\,dx = \left(-\frac{1}{3}\right) x^{-4/3}\,dx$

$$= -\frac{1}{3x^{4/3}}\,dx$$

(b) Given $f(x) = x^{-1/3}$, we wish to approximate $f(7.952)$, i.e., $f(8 - 0.048)$. Using the result above, with the approximation $f(x + \Delta x) \approx f(x) + df$, we obtain:

$$f(7.952) = f(8 - 0.048) \approx f(8) + df$$

$$= 0.5 + \frac{-1}{3(8^{4/3})}(-0.048)$$

$$= 0.5 + \frac{0.048}{48} = 0.5 + 0.001$$

$$= 0.501$$

THE MEAN-VALUE PRINCIPLE

4-10

Find the average value of sin(x) for $x \in [0,\pi]$.

****

The average value of f is $\frac{1}{\pi}\int_0^{\pi} \sin x \, dx =$

$$-\frac{1}{\pi}\left\{ \cos x \Big|_0^{\pi} \right\} = -\frac{1}{\pi}\{-1-1\} = \frac{2}{\pi}.$$

4-11

Find the sum of the values satisfying the Mean Value Theorem of $f(x) = \frac{x^3}{3} - 2x^2 + 3x + 4$ where $0 \le x \le 4$.

****

From the Mean Value Theorem we have that $f(4) - f(0) = f'(c)(4-0) \Rightarrow \frac{16}{3} - 4 = f'(c)(4) \Rightarrow$

$$\frac{4}{3} - 1 = f'(c) \Rightarrow \frac{1}{3} = c^2 - 4c + 3 \Rightarrow c^2 - 4c + \frac{8}{3} = 0.$$

The roots of $c^2 - 4c + 8/3 = 0$ are the values we are looking for. From the Quadratic formula we have

$$c = \frac{4 \pm \sqrt{16 - 4(1)(8/3)}}{2}.$$

The sum of these roots is 4.

4-12

If 1) f is continuous on $[2,4]$,

2) f' exists on (2,4),

3) $f'(c) = \frac{f(4) - f(2)}{2}$,

then $c \in (2,4)$. True or false?

****

False. The Mean Value Theorem states that under conditions stated in 1) and 2) there will be some c between 2 and 4 satisfying the condition stated in 3). It does not guarantee that there is no c outside of the interval (2,4) satisfying the condition stated in 3).

4-13

What does the MVT (Mean Value Theorem) guarantee for the given function on the given interval?

a. $f(x) = x^2 - 2x + 5$ on $[1,4]$. b. $g(x) = \frac{8}{(x-2)^2}$ on $[1,4]$.

****

a. $\frac{f(4)-f(1)}{4-1} = \frac{13-4}{3} = 3$, so the MVT assures us that there is a number c, $1 < c < 4$, such that $f'(c) = 3$.

b. NOTHING, because g is not continuous on the interval $[1,4]$.

4-14

What value(s) c (if any) are predicted by the Mean-Value theorem for the function $f(x) = (x-2)^3$ on the interval $[0,2]$?

****

Since $f(x) = (x-2)^3$ is everywhere differentiable (and continuous), the Mean Value theorem applies to $f(x)$ on the interval $[0,2]$. Therefore, there is at least one c such that:

1) $0 < c < 2$ and 2) $f'(c) = \dfrac{f(2) - f(0)}{2-0} = \dfrac{0+8}{2}$

$$= 4$$

$$f'(c) = 3(c-2)^2 = 4$$

$$(c-2)^2 = 4/3, \text{ so } c = 2 \pm \frac{2}{\sqrt{3}}$$

Since $2 + \frac{2}{\sqrt{3}} > 2$, this number does not satisfy condition (1) for c. Hence it is not a correct answer. Hence the desired c must be $c = 2 - \frac{2}{\sqrt{3}}$

Check: Note that indeed $2 - \frac{2}{\sqrt{3}}$ lies in the interior of $[0,2]$ since:

clearly $0 < \frac{2}{\sqrt{3}} < 2$, so $0 < 2 - \frac{2}{\sqrt{3}} < 2$

MAXIMUM AND MINIMUM VALUES OF A FUNCTION

4-15

Find the maximum and minimum values of $x^3 - 9x + 8$ on the interval $[-3,1]$, if they exist.

****

$f(x) = x^3 - 9x + 8 \qquad f'(x) = 3x^2 - 9 = 3(x^2 - 3)$

$f'(x) = 0$ at $x = \pm\sqrt{3}$, so the only critical point in $[-3, 1]$ is at $x = -\sqrt{3}$.

Check value of $f(x)$ at critical point and endpoints:

$$f(-3) = -27 + 27 + 8 = 8$$

$$f(-\sqrt{3}) = -3\sqrt{3} + 9\sqrt{3} + 8 = 8 + 6\sqrt{3}$$

$$f(1) = 1 - 9 + 8 = 0$$

Maximum $= 8 + 6\sqrt{3}$ Minimum $= 0$

4-16

Let $f(x) = x^{1/2}(1-x)$ for $x \geq 0$. Find the absolute maximum of $f(x)$ on the interval [0,4].

****

$f(x) = x^{1/2}(1-x) = x^{1/2} - x^{3/2}$

So, $f'(x) = \frac{1}{2}x^{-1/2} - \frac{3}{2}x^{1/2} = \frac{1}{2}\left(\frac{1}{\sqrt{x}} - 3\sqrt{x}\right)$

$= \frac{1}{2}\left(\frac{1-3x}{\sqrt{x}}\right)$

Note $f'(x)$ exists for all $x > 0$, so $f(x)$ is differentiable everywhere in the interior of [0,4]. Hence, the extremums of $f(x)$ occur either at a point where $f'(x) = 0$ or at the endpoints of the interval ($x = 0$ or $x = 4$)

$f'(x) = \frac{1}{2}\left(\frac{1-3x}{\sqrt{x}}\right) = 0$ precisely when $1 - 3x = 0$ or $x = 1/3$

Evaluating $f(x)$ at $x = 0$, $x = 1/3$, $x = 4$ we get:

$x = 0$: $f(x) = 0$

$x = 1/3$: $f(x) = \frac{1}{\sqrt{3}}\left(1 - \frac{1}{3}\right) = \frac{2}{3\sqrt{3}}$

$x = 4$: $f(x) = 2(1-4) = -6$

Hence: Max of $f(x)$ on $[0,4]$ is $\frac{2}{3\sqrt{3}}$ occuring at $x = 1/3$ (interior)

Min of $f(x)$ on $[0,4]$ is -6 occuring at $x = 4$ (end point)

4-17

Given that $f(x) = x^3 + ax^2 + bx$ has critical points at $x = 1$ and $x = 3$ find a and b and describe the points.

****

$f(x) = x^3 + ax^2 + bx$

Since the function has critical points at $x=1$ and at $x=3$ we know that $f'(1) = 0$

$f'(3) = 0$

$f'(x) = 3x^2 + 2ax + b$

Thus $f'(1) = 3 + 2a + b = 0$ or $2a + b = -3$

$f'(3) = 27 + 6a + b = 0$ or $6a + b = -27$

Solving the system
$$\begin{array}{r} 2a + b = -3 \\ 6a + b = -27 \\ \hline -4a \quad = 24 \end{array}$$
subtracting $\rightarrow a = -6$ hence $b = 9$.

Thus $f(x) = x^3 + -6x^2 + 9x$.

and $f'(x) = 3x^2 - 12x + 9 = 3(x^2 - 4x + 3) = 3(x-3)(x-1)$.

hence $f(x)$ has critical points at $x=1$ and $x=3$

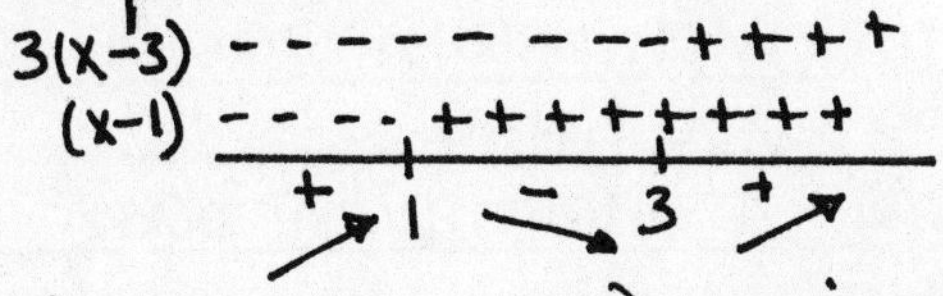

<u>Thus $f(x)$ has a maximum at $x=1$ and a minimum at $x=3$</u>

4-18

Use the second derivative test to identify the relative extreme points of

$f(x) = (x^2 - 6x + 8)^2$

****

$f(x) = (x^2-6x+8)^2$

we have critical points when $f'(x)=0$

so $f'(x) = 2(x^2-6x+8)^1 \cdot (2x-6) = 4(x-3)(x^2-6x+8)$

$f'(x) = 4(x-3)(x-4)(x-2) = 0$ when $x=2, x=3, x=4$.

To test for max. or min. points, find $f''(x)$

if $f''(x)>0$ at $x=a$ then the function is concave up ∪ and we have a minimum at $x=a$ if $f''(x)<0$ at $x=a$ then the function is concave down ∩ and we have a maximum at $x=a$. if $f''(x)=0$ the second derivative test fails.

Now $f''(x) = 4(x^2-6x+8) + 4(x-3)(2x-6)$
$= 4(x-4)(x-2) + 8(x-3)^2$

Test $f''(x)$ at

(a) $x=2 \rightarrow f''(2) = 4(-2)(0) + 8(1) = 8 > 0$, so $f(x)$ ∪ hence a min at $x=2$

(b) $x=3 \rightarrow f''(3) = 4(-1)(1) + 8(0) = -4 < 0$, so $f(x)$ ∩ hence a max at $x=3$

(c) $x=4 \rightarrow f''(4) = 4(0)(2) + 8(1) = 8 > 0$, so $f(x)$ ∪ hence a min at $x=4$

<u>So $f(x)$ has minimum points at $(2,0)$ and $(4,0)$</u>
since $f(2)=0$ and $f(4)=0$
<u>and $f(x)$ has a maximum point at $(3,1)$</u>
since $f(3)=1$

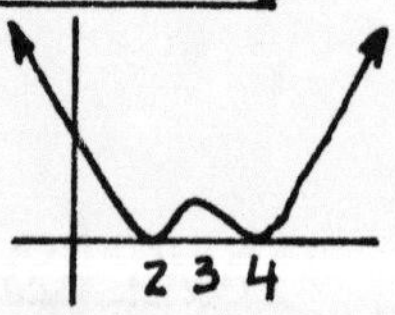

4-19

For the function f(x), find the maximum and minimum values on the interval [-1,5].

$$f(x) = \begin{cases} x^2 - 4 & \text{if } x \le 2 \\ x^2 - 8x + 12 & \text{if } x > 2 \end{cases}$$

****

$$f'(x) = \begin{cases} 2x & \text{if } x < 2 \\ 2x-8 & \text{if } x > 2 \end{cases} \qquad \begin{array}{l} f'(x) = 0 \text{ at } x = 0, 4 \\ f'(x) \text{ undefined at } x = 2 \end{array}$$

$f(-1) = -3$, $f(0) = -4$, $f(2) = 0$, $f(4) = -4$, $f(5) = -3$

The maximum value is 0 and it occurs at $x = 2$.
The minimum value is -4 and it occurs at $x = 0, 4$.

MAXIMUM-MINIMUM WORD PROBLEMS

4-20

Prove that the rectangular box with square base and top having a fixed surface area and maximum volume is a cube.

****

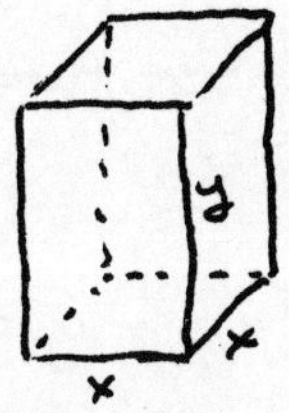

Let the dimensions of the box be x by x by y. (Since the base is square, the length and width are equal.) The two basic equations are

$$V = x^2 y \quad \text{and} \quad S = 2x^2 + 4xy,$$

with S, the surface area, being constant. Implicit differentiation can be used to obtain the result.

$$S = 2x^2 + 4xy \Rightarrow 0 = \frac{dS}{dx} = 4x + 4x\frac{dy}{dx} + 4y,$$

$$0 = x + y + x\frac{dy}{dx}, \quad \frac{dy}{dx} = -\frac{x+y}{x}$$

$$V = x^2 y \Rightarrow \frac{dV}{dx} = x^2\frac{dy}{dx} + 2xy = x^2\left(-\frac{x+y}{x}\right) + 2xy$$

$$= -x(x+y) + 2xy = -x(x-y)$$

Setting $\frac{dV}{dx} = 0$, we obtain $-x(x-y) = 0$, so $x = 0$ or $x = y$. $x = 0$ is impossible, for then the box would have 0 volume. Hence $x = y$, which implies the box is a cube.

The solution can also be obtained directly. Solve

$$S = 2x^2 + 4xy \text{ to obtain } y = \frac{S - 2x^2}{4x}.$$

Substitute into $V = x^2 y$ and differentiate.

$$V = x^2\left(\frac{S - 2x^2}{4x}\right) = \frac{1}{4}(xS - 2x^3)$$

Remember that S is constant.

$$\frac{dV}{dx} = \frac{1}{4}(S - 6x^2).$$

Setting $\frac{dV}{dx} = 0$, $S - 6x^2 = 0$, or $x = \sqrt{\frac{S}{6}}$.

Substitute in $y = \frac{S - 2x^2}{4x}$ to obtain

$$y = \frac{S - 2\left(\frac{S}{6}\right)}{4\sqrt{\frac{S}{6}}} = \sqrt{\frac{S}{6}}$$

Hence $y = \sqrt{\frac{S}{6}} = x$ and the box is a cube.

4-21

Acme Company's daily profit is given by the following equation

$$P(x) = -.03x^3 + 36x + 500, \quad x \geq 0$$

where x is the number of units sold each day and P(x) is the daily profit in dollars. Maximize Acme's daily profit.

****

$P'(x) = -.09x^2 + 36$ and $P''(x) = -.18x$

Set $P'(x) = 0$ and solve.

$$-.09x^2 + 36 = 0$$

$$-.09x^2 = -36$$

$$x^2 = 400$$

$x = 20$ or $x = -20$. Since x must be greater than or equal to zero, the only solution is $x = 20$. $P''(20) = -.18(20) = -3.6$

Since $P''(20) < 0$, $P(20)$ is the maximum daily profit.

$P(0) = 500$ and $P(20) = 980$, so \$980 is the maximum daily profit when 20 units are sold each day.

4-22

A cardboard box of 32 in^3 volume with a square base and open top is to be constructed. Find the minimum area of cardboard needed. (Neglect waste.)

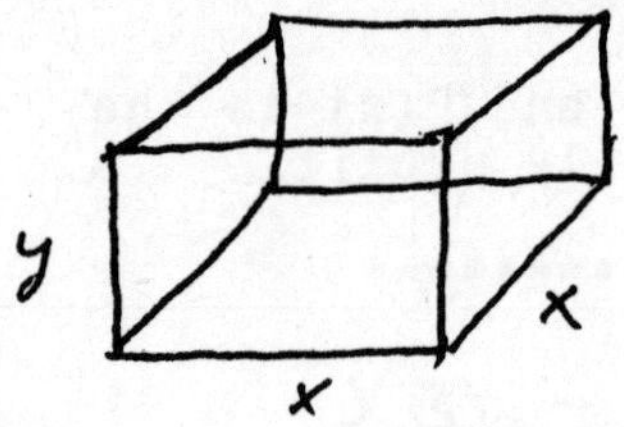

$$V = x^2 y = 32$$

$$A = x^2 + 4xy \quad \text{(OPEN TOP)}$$

$$x^2 y = 32 \Rightarrow y = 32/x^2 = 32x^{-2}$$

THUS

$$A = x^2 + 128x^{-1}$$

$$\frac{dA}{dx} = 2x - 128x^{-2}$$

$$= 2x - \frac{128}{x^2}$$

$\frac{dA}{dx} = 0$ if AND ONLY if $2x - \frac{128}{x^2} = 0$

i.e., $\frac{2x^3 - 128}{x^2} = 0$

$$\frac{2(x^3 - 64)}{x^2} = 0$$

OR $x^3 = 64.$

SO $x = 4.$

THUS

$$A = 16 + \frac{128}{4} = 48 \text{ in}^2$$

4-23

Farmer Brown wants to fence in a rectangular plot in a large field, using a rock wall which is already there as the north boundary. The fencing for the east and west sides of the plot will cost $3 a yard, but she needs to use special fencing which costs $5 a yard on the south side of the plot. If the area of the plot is to be 600 square yards, find the dimensions for the plot which will minimize the cost of the fencing.

****

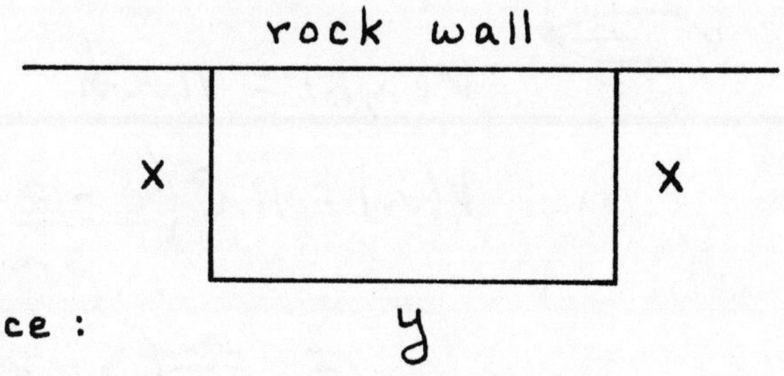

$xy = \text{area} = 600$

so $y = 600/x$

Minimize cost of fence:

$\text{Cost} = C = 2x(3) + y(5) = 6x + 5y$

substituting $y = 600/x$, $\quad C = 6x + \dfrac{3000}{x}$

Note that the possible range of x values is $x > 0$.

$$C' = 6 - \frac{3000}{x^2} = \frac{6x^2 - 3000}{x^2} = \frac{6}{x^2}(x^2 - 500)$$

So $C' = 0$ at $x = \sqrt{500} = 10\sqrt{5}$, and by the sign of C', it is a minimum of C.

$\therefore\ x = 10\sqrt{5}$ yards, $\quad y = \dfrac{600}{10\sqrt{5}} = 12\sqrt{5}$ yards.

4-24

A container with a closed top is to be constructed into a right circular cylinder. If the surface area is fixed find the ratio of the height to the radius which will maximize the volume.

****

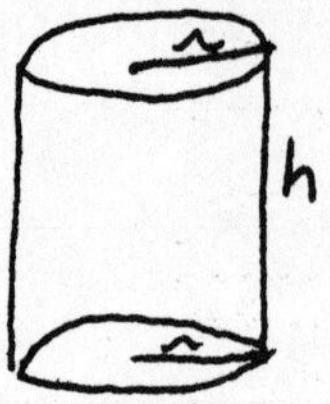

Volume $= \pi r^2 h$, S = surface area

$V = \pi r^2 h$, $S = 2r\pi h + 2r^2\pi$

$$\frac{S - 2r^2\pi}{2r\pi} = h$$

$$V(r,h) = \pi r^2 h$$

$$V(r) = \pi r^2\left(\frac{S - 2r^2\pi}{2r\pi}\right)$$

$$= \frac{r}{2}(S - 2r^2\pi) = \frac{S}{2}r - \pi r^3$$

Now $V'(r) = \frac{S}{2} - 3\pi r^2$. Set $V'(r) = 0 \Rightarrow \frac{S}{2} = 3\pi r^2 \Rightarrow$ $r = \sqrt{\frac{S}{6\pi}}$. From the 2nd derivative test we have that $V''\left(\sqrt{\frac{S}{6\pi}}\right) = -\sqrt{6\pi S} < 0 \Rightarrow$ that when $r = \sqrt{\frac{S}{6\pi}}$ we have a maximum.

Now we must find h. $h = \left.\frac{S - 2r^2\pi}{2r\pi}\right|_{r=\sqrt{\frac{S}{6\pi}}} \Rightarrow$

$$h = \frac{S - 2\left(\frac{S}{6\pi}\right)\pi}{2\sqrt{\frac{S}{6\pi}}\,\pi} = \frac{2S}{6\sqrt{\frac{S}{6\pi}}\,\pi} = 2\sqrt{\frac{S}{6\pi}} = 2r.$$

The ratio of height to the radius is $2r : r$ $\Rightarrow 2:1$.

4-25

A gas pipeline is to be constructed from a storage tank to a house which sits 600 feet down a road and 300 feet back from the road (see picture). Pipe laid along the road costs $8.00 per foot, while pipe laid off the road costs $10.00 per foot. What is the minimum cost for which this pipeline can be built? (Assume the pipeline path is piecewise linear, with at most two pieces.)

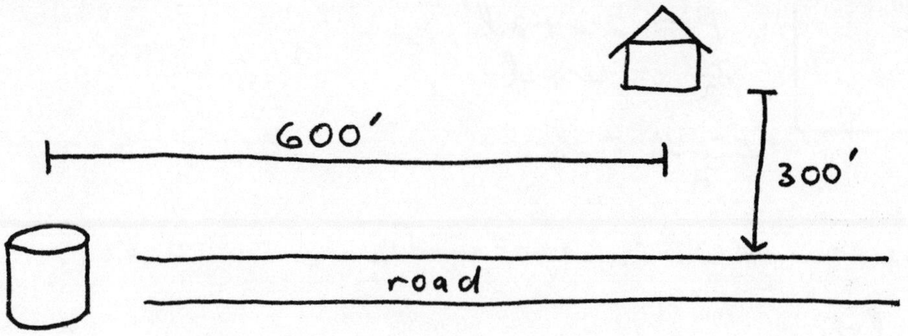

****

Let x represent the distance from where the pipeline will leave the road to the point 600 feet down the road from the tank.

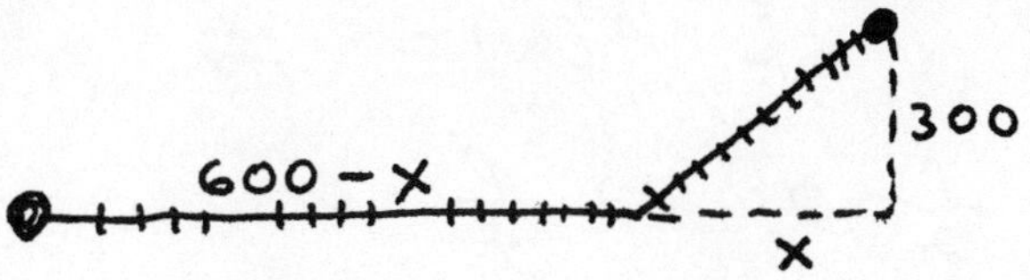

$0 \le x \le 600$

$$\text{Cost} = C(x) = 8(600-x) + 10\sqrt{x^2+300^2}$$

$$C'(x) = -8 + \frac{10x}{\sqrt{x^2+300^2}} = 0$$

$$\Rightarrow 10x = 8\sqrt{x^2+300^2} \Rightarrow 100x^2 = 64(x^2+300^2)$$

$$\Rightarrow 36x^2 = 64(300^2) \Rightarrow x^2 = \frac{8^2\,300^2}{6^2}$$

$$\Rightarrow x = \pm\frac{8(300)}{6} = \oplus 400 \quad (-400 \text{ is extraneous})$$

$$C(0) = 7800, \quad C(600) = 3000\sqrt{5} \approx 6708, \quad C(400) = 6600$$

∴ \$6600 is the minimum cost, obtained by using 200 ft. of road pipe and 500 ft. of off-road pipe.

4-26

Let R = {z|z is the area of a rectangle with a fixed perimeter p}. Show that the largest value of R is $\frac{p^2}{16}$.

****

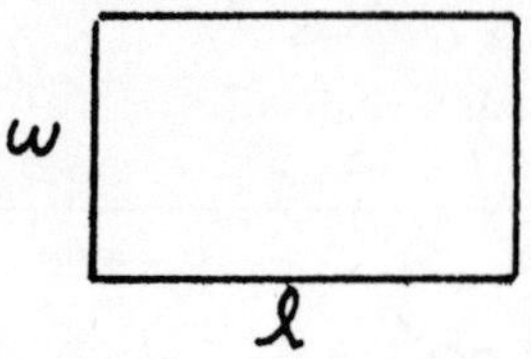

$$p = 2w + 2l$$
$$\frac{p}{2} = w + l$$
$$\frac{p-2w}{2} = l$$

The area of the rectangle is $A = lw$.

$$A = lw$$
$$= \left(\frac{p-2w}{2}\right)w = \frac{p}{2}w - w^2.$$

Now we have the area in terms of w i.e.

$$A(w) = \frac{p}{2}w - w^2.$$

$A'(w) = \frac{p}{2} - 2w$. Set $A'(w) = 0 \Rightarrow$

$$\frac{p}{2} - 2w = 0 \Rightarrow \frac{p}{2} = 2w \Rightarrow \frac{p}{4} = w.$$

When $w = p/4$ then $l = \frac{p - 2(p/4)}{2} \Rightarrow$

$l = \frac{p - p/2}{2} = \frac{p/2}{2} = p/4$. Hence we have an extrema when $w = p/4$ and $l = p/4$ and the area is $\frac{p^2}{16}$.

4-27

A small, excellent, liberal arts college in central Michigan is planning to build a new, 18000-cubic-foot outdoor swimming pool with rectangular concrete sides. The square bottom is to boast a majestic portrait in mosaic tile of the distinguished Dr. J. F. Putz (and is expected to generate a robust helicopter business from tourists and curiosity seekers clambering to see it from the air). The bottom is to be at least 10 feet on each side and at most 40 feet on each side. If the cost of the concrete to be used for the sides is \$15 per square foot and the cost of the tile to be used for the bottom is \$20 per square foot, what should the dimensions be in order to minimize the cost of materials for this worthy and exciting endeavor?

****

LET THE VARIABLES BE CHOSEN AS SHOWN IN THE FIGURE AND LET $C(x)$ BE THE COST OF THE MATERIALS.

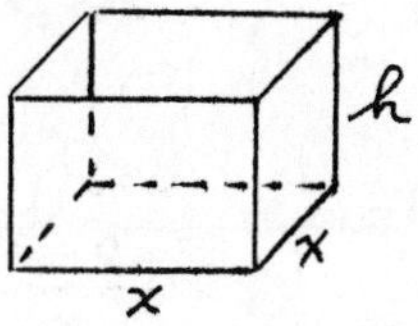

$C(x) = 15(4xh) + 20x^2$

BUT WE KNOW $x^2h = 18{,}000$, SO $h = \dfrac{18000}{x^2}$.

SO $C(x) = 60x\,\dfrac{18000}{x^2} + 20x^2 = 108\cdot10^4x^{-1} + 20x^2$

WITH $10 \le x \le 40$.

$C'(x) = -108\cdot10^4x^{-2} + 40x = 0$

$40x = 108\cdot10^4x^{-2}$

$x^3 = 27\cdot10^3$

$x = 30$

THE POSSIBLE MINIMUM POINTS ARE $x=30$, AND THE ENDPOINTS $x = 10$ AND $x = 40$.

$C(10) = 108\cdot10^4\cdot10^{-1} + 20\cdot10^2 = 108{,}000 + 2{,}000 = 110{,}000$

$C(30) = 108\cdot10^4\cdot30^{-1} + 20\cdot30^2 = 36{,}000 + 18{,}000 = 54{,}000$

$C(40) = 108\cdot10^4\cdot40^{-1} + 20\cdot40^2 = 27{,}000 + 32{,}000 = 59{,}000$

THEREFORE, $x = 30$ MINIMIZES THE COST.

$h = \dfrac{18000}{30^2} = 20$

SO THE DIMENSIONS WHICH MINIMIZE THE COST ARE $30' \times 30' \times 20'$.

4-28

A plastic right cylinder with closed ends is to hold V cubic feet. If there is no waste in construction, find the ratio between the height and diameter that results in the minimum use of material?

****

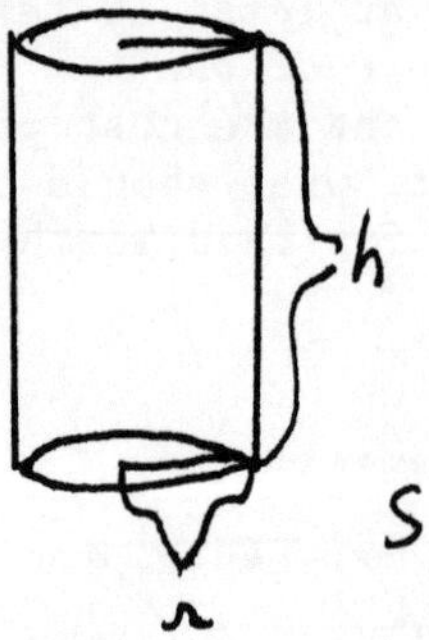

$$V = \pi r^2 h \Rightarrow h = \frac{V}{\pi r^2}$$

$$S = 2\pi r^2 + 2\pi r h \text{ (surface area)}$$

$$S(r,h) = 2\pi r^2 + 2\pi r h$$

$$S(r) = 2\pi r^2 + 2\pi r\left(\frac{V}{\pi r^2}\right)$$

$$= 2\pi r^2 + 2\pi V r^{-1}$$

Since we now have S in terms of a single variable we can find S' & S''.

$$S'(r) = 4\pi r - 2\pi V r^{-2} = 2\pi r^{-2}(2r^3 - V).$$

$$S''(r) = 4\pi + 4\pi V r^{-3} = 4\pi\left(1 + \frac{V}{r^3}\right).$$

From $S'(r)$ we get 0 and $\left(\frac{V}{2}\right)^{1/3}$ as critical numbers. When $r = 0$ we have no volume and $S''\left(\sqrt[3]{\frac{V}{2}}\right) = 12\pi > 0 \Rightarrow r = \left(\frac{V}{2}\right)^{1/3}$ is a minimum.

Now $h = \frac{V}{\pi r^2}\Big|_{r=\sqrt[3]{\frac{V}{2}}} = \frac{V}{\pi \frac{V^{2/3}}{2^{2/3}}} = \frac{2^{2/3} V^{1/3}}{\pi}$. The ratio of height and diameter is $\frac{h}{2r} = \frac{\frac{2^{2/3} V^{1/3}}{\pi}}{2^{2/3} V^{1/3}} = \frac{1}{\pi}$. $\therefore$ the ratio of height to diameter is $1 : \pi$.

4-29

A right circular cone is inscribed in a hemisphere of radius 2 as shown. Find the dimensions of one such cone with maximum volume.

[Note: The volume of a cone is given by $V = \frac{1}{3}\pi r^2 h$]

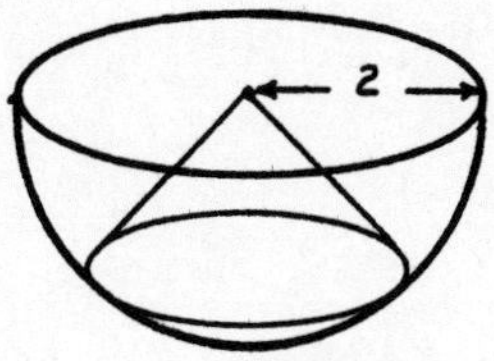

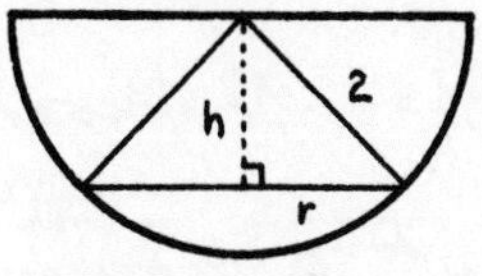

Cross Section

****

Note from the cross section that

$$h^2 + r^2 = 2^2 \quad \text{so} \quad r^2 = 4 - h^2$$

Since $V = \frac{1}{3}\pi r^2 h$ we have $V = \frac{1}{3}\pi (4 - h^2) h$

Thus $\frac{dV}{dh} = \frac{\pi}{3}(4 - 3h^2)$.

Now $\frac{dV}{dh} = 0$ when $\frac{\pi}{3}(4 - 3h^2) = 0 \Rightarrow h = \sqrt{4/3}$

Also $\frac{d^2V}{dh^2} = -2\pi h$ which is < 0 for $h = \sqrt{4/3}$

So by the second derivative test, $h = \sqrt{4/3}$ gives a maximum value for V.

The required dimensions are

$$h = \sqrt{\frac{4}{3}} = \frac{2\sqrt{3}}{3} \quad \text{and} \quad r = \sqrt{\frac{8}{3}} = \frac{2\sqrt{6}}{3}$$

The maximum volume is: $\frac{16\pi\sqrt{3}}{27}$ cubic units

4-30

Find two positive numbers whose sum is 8 such that when the cube of the first number is multiplied by the second number the result is a maximum.

****

Let x = the first number

so $8-x$ = the second number

We want to maximize y, where

$$y = x^3(8-x) = 8x^3 - x^4 \quad \text{and } 0 \le x \le 8.$$

so $\dfrac{dy}{dx} = 24x^2 - 4x^3$. To find a maximum, we set equal to 0. Thus...

$$24x^2 - 4x^3 = 0$$
$$4x^2(6-x) = 0$$

so 0 and 6 are the critical numbers.

If $x=0$, $y=0$ which is a minimum (not wanted)

If $x=6$, $y=432$ and it is clear that this is a maximum.

Thus the first number is 6 and the second is 2.

4-31

A rectangular area of 1050 square feet is to be enclosed by a fence, then divided down the middle by another piece of fence. The fence down the middle costs $0.50 per running foot, and the other fence costs $1.50 per running foot. Find the minimum possible cost for the fence.

****

LET x = WIDTH

y = LENGTH

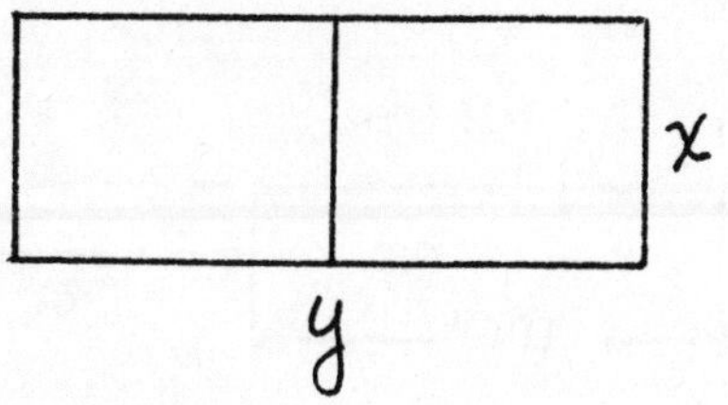

$$xy = 1050$$

$$y = \frac{1050}{x}$$

$$C = 1.50\left(2x + 2\left(\frac{1050}{x}\right)\right) + 0.50x$$

$$C = 3.5x + \frac{3150}{x}$$

$$\frac{dC}{dx} = 3.5 - \frac{3150}{x^2}$$

$$3.5 - \frac{3150}{x^2} = 0$$

$$x = 30$$

$$\frac{d^2C}{dx^2} = \frac{6300}{x^3}$$

$$\left.\frac{d^2C}{dx^2}\right|_{x=30} = +$$

THUS, $x = 30$ LEADS TO A MINIMUM FOR C

$$C = 3.5(30) + \frac{3150}{30}$$

$$C = \$210$$

4-32

We have nine hundred feet of fence and wish to fence in a rectangular area, using a one hundred foot building wall as part of the boundary (see diagram).

a. What is the range of possible x-values?

b. What value of x gives the largest area?

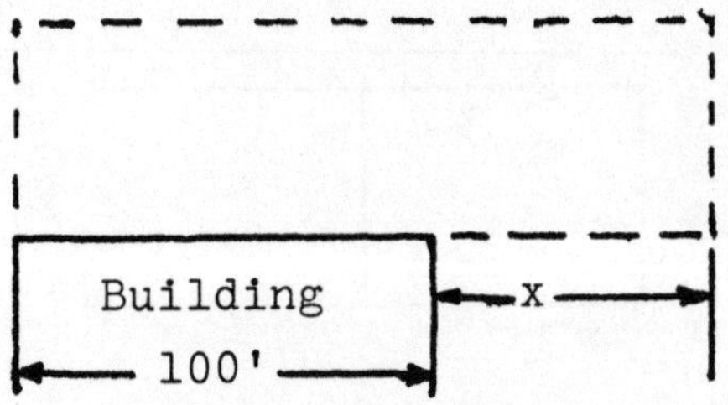

****

a. $0 \le x \le 400$.

b. base of rectangle $= 100 + x$

height of rectangle $= \frac{1}{2}(900 - x - (100 + x)) = \frac{1}{2}(800 - 2x)$
$= 400 - x$

Area $= A(x) = (100 + x)(400 - x) = 40000 + 300x - x^2$.

The maximum occurs at $x = 0$ or $x = 400$ or at a critical point of $A(x)$ in $(0, 400)$.

$A'(x) = 300 - 2x$.

Set $A'(x) = 0$: $300 - 2x = 0$; $300 = 2x$; $x = 150$.

Points to check

$x = 0$	Area $= 100 \cdot 400 = 40000$ sq. feet.	
$x = 150$	Area $= 250 \cdot 250 = 62500$ sq. feet.	
$x = 400$	Area $= 500 \cdot 0 = 0$ sq. feet.	

The maximum occurs at $x = 150$.

4-33

A square is to be cut from each corner of a piece of paper which is 8 cm. by 10 cm. and the sides are to be folded up to create an open box. What should the side of the square be for maximum volume?

****

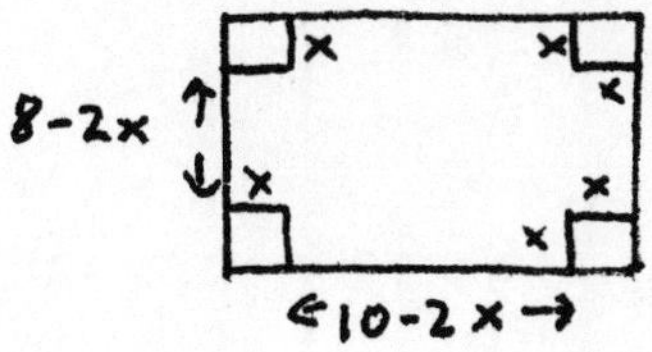

$$V_{box} = l \cdot w \cdot h$$

$$V_{box} = (10-2x)(8-2x)x = 80x - 36x^2 + 4x^3$$

Maximum volume will occur where $\frac{dV}{dx} = 0$

$$\frac{dV}{dx} = 80 - 72x + 12x^2 = 0$$

$$x = 3 + \frac{\sqrt{21}}{3} \text{ or } 3 - \frac{\sqrt{21}}{3}$$

To determine which of these will give <u>maximum</u> volume, use 2nd derivative test.

$$\frac{d^2V}{dx^2} = -72 + 24x$$

If $x = 3 + \frac{\sqrt{21}}{3}$ this is $-72 + 24\left(3 + \frac{\sqrt{21}}{3}\right) = 8\sqrt{21}$ POS. MIN.

If $x = 3 - \frac{\sqrt{21}}{3}$ this is $-72 + 24\left(3 - \frac{\sqrt{21}}{3}\right) = -8\sqrt{21}$ NEG. MAX

Answer is $3 - \frac{\sqrt{21}}{3} \approx 1.47$ cm

4-34

A carpenter wishes to build a bin with lid in the corner of a room, utilizing the corner walls and floor. What dimensions will hold 8 cubic feet using the least plywood, if the base is square?

****

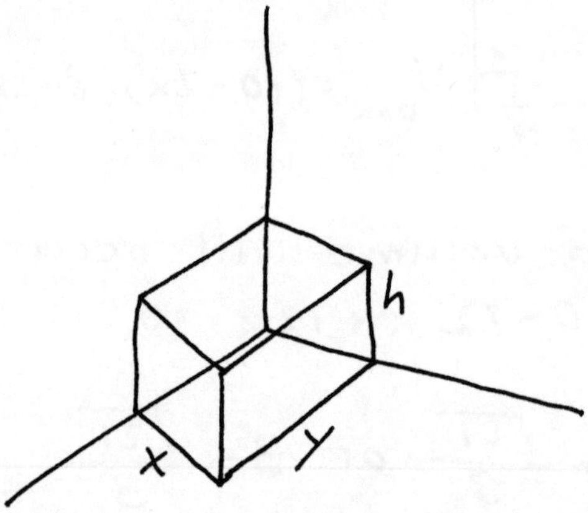

The plywood required is $P = xh + yh + xy = 2xh + x^2$ since the base is square. Since $V = 8 = xyh$ and $x = y$, $x^2h = 8$. Thus $h = 8/x^2$ and $P = x^2 + 16/x$. $P' = 2x - 16/x^2 = 0$ when $2x^3 = 16$, $x = 2$. Thus the most economical bin is a $2 \times 2 \times 2$ cube.

4-35

Consider all rectangles of the type shown in the figure for $0 < t < 1$. What value(s) of t (if any) will yield the rectangles of largest and smallest area?

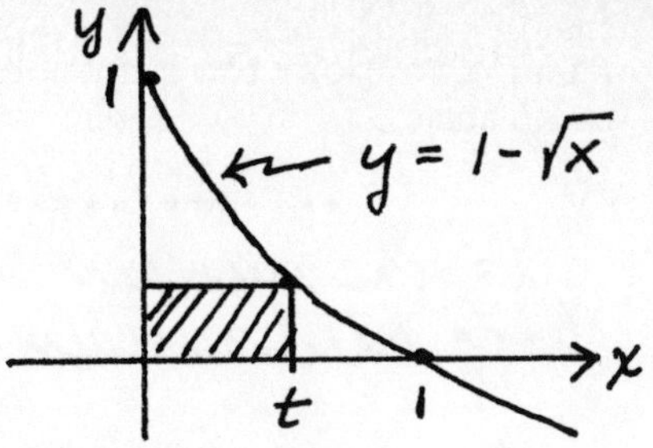

****

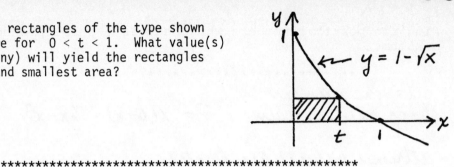

The area A of a typical rectangle is given by:

$$A = t(1-\sqrt{t}) = t - t^{3/2}$$

To find the extreme values of A, we set the derivative equal to 0:

$$\frac{dA}{dt} = 1 - \frac{3}{2}t^{1/2} = 0 \quad \text{when} \quad \frac{3}{2}t^{1/2} = 1$$

or $t^{1/2} = \frac{2}{3}$ or $t = 4/9$

So, an extreme value of A occurs when $t = \frac{4}{9}$ and it is intuitively clear that this gives a maximum of A. To check this, use the second derivative test:

$$\frac{d^2A}{dt^2} = -\frac{3}{4}t^{-1/2} = -\frac{3}{4\sqrt{t}}. \quad \text{Hence} \quad \frac{d^2A}{dt^2} < 0$$

for all $t > 0$, which confirms that:

$t = 4/9$ gives a maximum of A

Since the degenerate rectangles with area 0 that occur for $t = 0$ and $t = 1$ are excluded by hypothesis ($0 < t < 1$):

there is no rectangle of minimum area.

4-36

Find two positive numbers whose sum is 6 whose product is a maximum.

****

x = one number

$6-x$ = other number. $P = x(6-x) = 6x - x^2$

extreme values are located where 1st derivative = 0

$P' = 6 - 2x$

$P' = 0 \Rightarrow 6 - 2x = 0 \Rightarrow x = 3$

Use 2nd derivative test to determine maximum or minimum.

$P'' = -2$

$P''(3) = -2$ negative value $\Rightarrow x = 3$ represents a maximum

so two numbers are $x = 3$ and $6 - x = 3$.

THE FIRST DERIVATIVE TEST

4-37

Find all relative maximums and minimums of $f(x) = 2x^3 - 3x^2 - 12x + 5$ using the first derivative test.

****

Possible relative extremums exist where $f'(x) = 0$.
$f'(x) = 6x^2 - 6x - 12 = 6(x^2 - x - 2) = 6(x-2)(x+1)$, so $f'(x) = 0$ when $x = 2$ or $x = -1$.

Choosing a value less than $x = -1$, say $x = -2$, we see that $f'(-2) = 24 + 12 - 12 = 24 > 0$, which means $f(x)$ is increasing for $x < -1$. Choosing a value between $x = -1$ and $x = 2$, say $x = 0$, we see that $f'(0) = 0 - 0 - 12 = -12 < 0$, which means that $f(x)$ is decreasing for $-1 < x < 2$. Choosing a value of x greater than $x = 2$, say $x = 3$, we see that $f'(3) = 54 - 18 - 12 = 24 > 0$, which means $f(x)$ is increasing for $x > 2$.

Conclusion: Since $f(x)$ changes from increasing to decreasing at $x = -1$, a relative maximum exists at $x = -1$, and is the point $(-1, 12)$.

Since $f(x)$ changes from decreasing to increasing at $x = 2$, a relative minimum exists at $x = 2$, and is the point $(2, -15)$.

4-38

Let $f(x)=\frac{1}{3}x^3 - x$. Use the first derivative test to determine the following: (a) Where f(x) is increasing and (b) where f(x) is decreasing.

****

(a) $f'(x) = x^2-1$. Since f is increasing when $f'(x)>0$, let $f'(x)>0$. Then $x^2-1>0$ and $(x+1)(x-1)>0$. Hence either

$x+1>0$ and $x-1>0$
$x>-1$ and $x>1$
from which we get $x>1$

OR

$x+1<0$ and $x-1<0$
$x<-1$ and $x<1$
from which we get $x<-1$

Hence f is increasing when $x>1$ when $x<-1$.

(b) Since f is decreasing when $f'(x)<0$, let $f'(x)<0$ and solve the resulting inequality.
Then $x^2-1<0$ so that $(x+1)(x-1)<0$.
Hence either

$x+1<0$ and $(x-1)>0$
$x<-1$ and $x>1$
from which we get __No__ Solution

OR

$x+1>0$ and $x-1<0$
$x>-1$ and $x<1$
from which we get $-1<x<1$.

So f is decreasing if $-1<x<1$.

Therefore f is increasing when $x>1$ and when $x<-1$ but decreasing when $-1<x<1$.

■■ **4-39**

Find and describe all local extrema of $f(x) = x^{5/3} - 5x^{2/3}$

****

$$f(x) = x^{5/3} - 5x^{2/3}$$

$$f'(x) = \tfrac{5}{3}x^{2/3} - \tfrac{10}{3}x^{-1/3} = \tfrac{5}{3}\left[x^{2/3} - \frac{2}{x^{1/3}}\right]$$

$$= \tfrac{5}{3}\left[\frac{x-2}{x^{1/3}}\right]$$

$$\text{so } f'(x) = \tfrac{5}{3}\left(\frac{x-2}{x^{1/3}}\right)$$

now we have critical points when $f'(x) = 0$ or when $f'(x)$ is undefined

firstly $f'(x) = 0$ when $\frac{x-2}{x^{1/3}} = 0 \rightarrow x-2 = 0$ when $x = 2$

secondly $f'(x)$ is undefined when $x^{1/3} = 0$, when $x = 0$

$\tfrac{5}{3}(x-2)$ ‑ ‑ ‑ ‑ ‑ ‑ ‑ ‑ ‑ ‑ ‑ ‑ ‑ + + + + + +

$x^{1/3}$ ‑ ‑ ‑ ‑ ‑ + + + + + + + + + + + + +

notice $f(0) = 0$ so at $(0,0)$ we have a vertical tangent, or 'elbow'

$f'(x)$ +   0   −   2   +

f(x) has a relative minimum at x=2, and a relative maximum at x=0

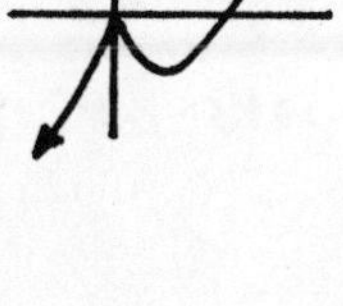

THE SECOND DERIVATIVE TEST

4-40

Let $f(x)=\frac{1}{3}x^3 - x$. Find the critical values of f and then use the second derivative to determine whether these critical values are relative max. or relative minimum.

****

$f'(x) = x^2 - 1$. So let $f'(x) = 0$ to get $x = \pm 1$ for critical values.
$f''(x) = 2x$ and since $f''(1) = 2 > 0$, the point $(1, -\frac{2}{3})$ is a relative minimum point. Since $f''(-1) = -2 < 0$, the point $(-1, \frac{2}{3})$ is a relative maximum point.

4-41

Find all relative maximums and minimums of $f(x) = x^4 - (4/3)x^3 - 12x^2 + 1$ using the second derivative test.

****

Possible relative extremums exist where $f'(x) = 0$.

$f'(x) = 4x^3 - 4x^2 - 24x = 4x(x^2 - x - 6) = 4x(x-3)(x+2)$,

so $f'(x) = 0$ when $x = 0$, $x = 3$, or $x = -2$.

$f''(x) = 12x^2 - 8x - 24$

$f''(0) = 0 - 0 - 24 = -24 < 0$, so a relative maximum exists at $x = 0$, and is the point $(0, 1)$.

$f''(3) = 108 - 24 - 24 = 60 > 0$, so a relative minimum exists at $x = 3$, and is the point $(3, -62)$.

$f''(-2) = 48 + 16 - 24 = 40 > 0$, so a relative minimum exists at $x = -2$, and is the point $(-2, -61/3)$.

RELATED RATES

4-42

A cylindrical can is undergoing a transformation in which the radius and height are varying continuously with time t. The radius is increasing at 4 in/min, while the height is decreasing at 10 in/min. Is the volume of the can increasing or decreasing, and at what rate, when the radius is 3 inches and the height is 5 inches?

****

Volume of can = $V = \pi r^2 h$

$$\frac{dV}{dt} = \pi\left[2rh\, dr/dt + r^2\, dh/dt\right]$$

$$= \pi\left[8rh - 10r^2\right], \text{ since } dr/dt = +4 \text{ in/min}$$

$$\text{and } dh/dt = -10 \text{ in/min}$$

$$\therefore \left.\frac{dV}{dt}\right|_{\substack{r=3\\h=5}} = \pi\left[8(3)(5) - 10(3^2)\right]$$

$$= \pi\left[120 - 90\right] = 30\pi$$

The volume of the can is <u>increasing</u> at the rate of 30π cubic inches per minute, when the radius is 3 inches and the height is 5 inches.

4-43

A frugal young man has decided to extract one of his teeth by tying a stout rubber band from his tooth to the chain on a garage door opener which runs on a horizontal track 3 feet above his mouth. If the garage door opener moves the chain at 1/4 ft/sec, how fast is the rubber band expanding when it is stretched to a length of 5 feet?

****

LET THE VARIABLES BE CHOSEN AS SHOWN IN ELABORATE FIGURE. WE KNOW THAT $y^2 = x^2 + 9$ AND THAT $\frac{dx}{dt} = 1/4$. WE NEED $\frac{dy}{dt}$ WHEN $y = 5$.

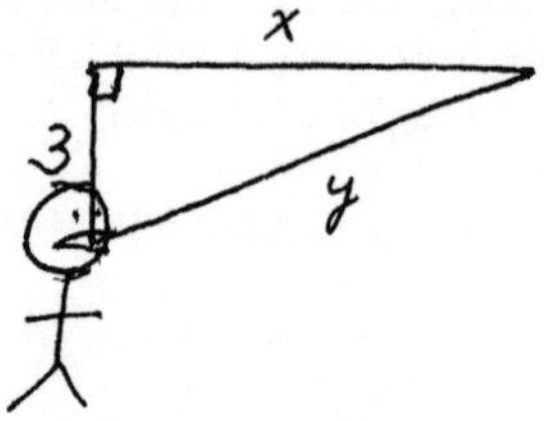

DIFFERENTIATING WITH RESPECT TO t, WE HAVE

$$2y\frac{dy}{dt} = 2x\frac{dx}{dt}$$

$$y\frac{dy}{dt} = x\frac{dx}{dt}.$$

WHEN $y = 5$, WE HAVE

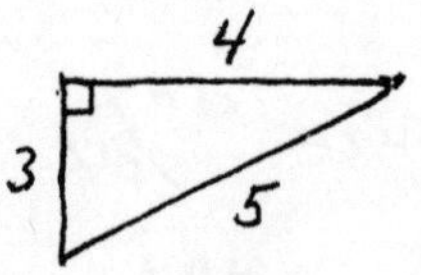

SO $5\frac{dy}{dt} = 4 \cdot \frac{dx}{dt} = 4 \cdot \frac{1}{4} = 1$

$$\frac{dy}{dt} = \frac{1}{5}.$$

THE RUBBER BAND IS EXPANDING AT $\frac{1}{5}$ FT/SEC.

4-44

Two straight roads intersect at right angles in Newtonville. Car A is on one road moving toward the intersection at a speed of 50 m.p.h. Car B is on the other road moving away from the intersection at a speed of 30 m.p.h. When car A is 2 miles from the intersection and car B is 4 miles from the intersection:

(a) How fast is the distance between the cars changing?

(b) Are the cars getting closer together or farther apart?

****

(a)

$$D^2 = x^2 + y^2$$

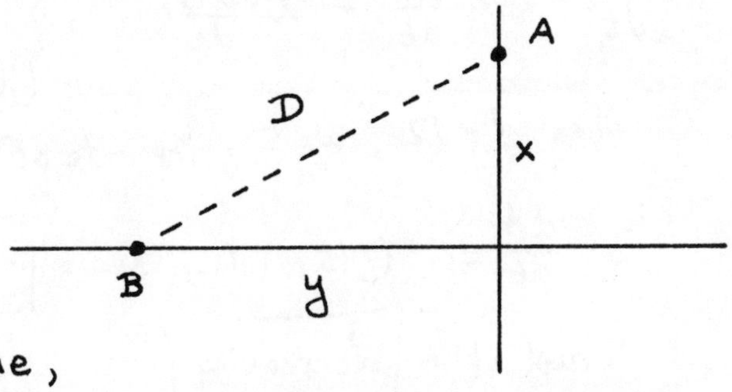

Differentiating with respect to time,

$$2DD' = 2xx' + 2yy', \quad \text{so} \quad D' = \frac{xx' + yy'}{D}$$

At the time in question,

$x = 2$, $y = 4$, $D = \sqrt{x^2 + y^2} = \sqrt{20}$, and

$x' = -50$ (x decreasing), $y' = +30$ (y increasing)

$$\therefore \quad D' = \frac{(2)(-50) + (4)(30)}{\sqrt{20}} = \sqrt{20} \text{ m.p.h.}$$

(b) $D' > 0$ so D is increasing and the cars are getting farther apart.

4-45

The length of a rectangle is increasing at the rate of 7 ft/sec, while the width is decreasing at the rate of 3 ft/sec. At one time, the length is 12 feet and the diagonal is 13 feet. At this time find the rate of change in
a) the area
b) the perimeter
and tell whether each is increasing or decreasing.

****

a) $A = l \cdot w$

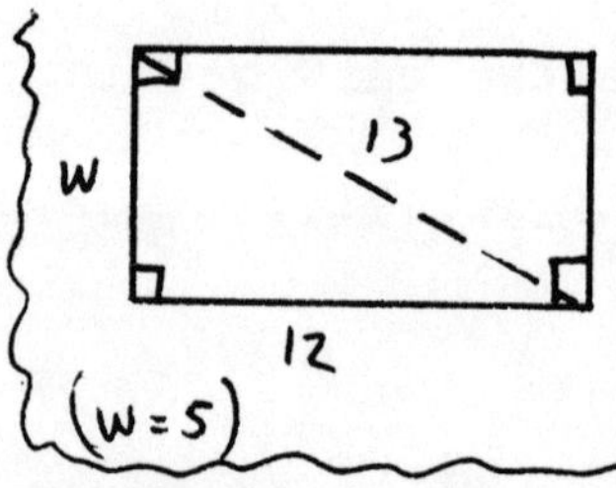

$\frac{dl}{dt} = 7$

$\frac{dw}{dt} = -3$

$$\frac{dA}{dt} = w \cdot \frac{dl}{dt} + l \cdot \frac{dw}{dt}$$

when $l = 12$, $w = 5$, the rate of change of the area is

$$\frac{dA}{dt} = (7)(5) + (12)(-3) = \boxed{-1 \text{ sq. ft/sec}}$$

and it is $\boxed{\text{decreasing}}$

b) $P = 2l + 2w$

$$\frac{dP}{dt} = 2 \cdot \frac{dl}{dt} + 2\frac{dw}{dt}$$

$$\frac{dP}{dt} = (2)(7) + (2)(-3) = \boxed{8 \text{ ft/sec}}$$

and it is $\boxed{\text{increasing}}$

4-46

A particle starts at the origin and moves along the parabola $y = x^2$ such that it distance from the origin increases at 4 units per second. How fast is its x-coordinate changing as it passes through the point (1,1) ?

****

Let s = distance of particle from (0,0)

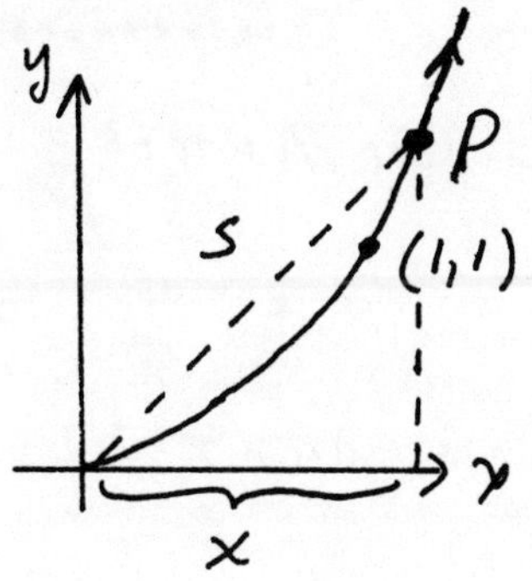

Given: $\frac{ds}{dt} = 4$

To find: $\left.\frac{dx}{dt}\right]_{x=1} = ?$

We need a relation between s and x

In general, $s^2 = x^2 + y^2$

Since the particle is restricted to $y = x^2$:

$$s^2 = x^2 + (x^2)^2 = x^2 + x^4$$

Differentiating both sides of the above equation with respect to t:

$$2s \cdot \frac{ds}{dt} = 2x \cdot \frac{dx}{dt} + 4x^3 \frac{dx}{dt}$$

or: $s \frac{ds}{dt} = x \cdot \frac{dx}{dt} + 2x^3 \cdot \frac{dx}{dt}$.

By hypothesis $ds/dt = 4$. Also, when $x = 1$, $s = \sqrt{2}$

Hence: $\sqrt{2} \cdot 4 = 1 \cdot \left.\frac{dx}{dt}\right]_{x=1} + 2 \cdot 1^3 \cdot \left.\frac{dx}{dt}\right]_{x=1}$

So: $\left.\frac{dx}{dt}\right]_{x=1} = \frac{4}{3}\sqrt{2}$ units / sec.

4-47

When a stone is dropped in a pool, a circular wave moves out from the point of impact at a rate of six inches per second. How fast is the area enclosed by the wave increasing when the wave is two inches in radius? (Recall: $A = \pi r^2$, where A = area, r = radius).

****

Given $A = \pi r^2$, we get $dA = \pi \cdot 2r \cdot dr$.

In the problem, $dr = 6$ and $r = 2$.

Thus $dA = \pi \cdot 2 \cdot 2 \cdot 6 = 24\pi$ sq. in./sec.

4-48

Suppose that $y = 2x^2 - 3x + 1$.

(a) Find and simplify a formula for the y increment, Δy.

(b) Find a formula for the y differential, dy.

****

(a) $\Delta y = 2(x+\Delta x)^2 - 3(x+\Delta x) + 1 - [2x^2 - 3x + 1]$

$= 2x^2 + 4x(\Delta x) + 2(\Delta x)^2 - 3x - 3(\Delta x) + 1 - 2x^2 + 3x - 1$

$= 4x(\Delta x) + 2(\Delta x)^2 - 3(\Delta x)$

(b) $dy = y'\,dx = (4x - 3)\,dx$

4-49

Using differentials, find an approximation for the cube root of 30.

****

Let $f(x) = \sqrt[3]{x} = x^{1/3}$. If Δx is small and nonzero, then $f(x+\Delta x) \approx f(x) + f'(x)\Delta x$.

$\sqrt[3]{30} = f(30) = f(27+3)$. Therefore $x = 27$ and $\Delta x = 3$. Since $f'(x) = \frac{1}{3}x^{-2/3} = \frac{1}{3x^{2/3}}$,

$$\sqrt[3]{30} = f(27+3) \approx f(27) + f'(27)(3)$$

$$\approx 3 + \frac{1}{3(9)} \cdot 3$$

$$\approx 3 + \tfrac{1}{9} = 3.\overline{1}$$

A calculator check shows $\sqrt[3]{30} \approx 3.1072$.

4-50

A particle moves along a path described by $y = x^2$. At what point along the curve are x and y changing at the same rate? Find this rate if at that time t we have $x = \sin(t)$ and $y = \sin^2 t$.

****

$\frac{dy}{dt} = 2x\frac{dx}{dt}$ $\quad \frac{dy}{dt} = \frac{dx}{dt} \Rightarrow \frac{dx}{dt} = 2x\frac{dx}{dt} \Rightarrow 1 = 2x$
$\Rightarrow \frac{1}{2} = x$

at $x = 1/2$, $y = (1/2)^2 = 1/4$ so point is $(1/2, 1/4)$.

now for $x = \sin t$, $\frac{dx}{dt} = \cos t$

at $x = 1/2$, $\sin t = 1/2 \Rightarrow t = \pi/6$ so $\frac{dx}{dt} = \cos \pi/6 = \sqrt{3}/2$

4-51

Two of the cutest little puppies you've ever seen begin running from the same point. One of them--a big-eyed little rascal with the endearing habit of cocking his head to one side--romps due east at 5 miles per hour. Frisky, though pudgy, the other darling little tyke gambols directly northward at only 4 miles per hour. At what rate is the distance between the puppies changing after 2 hours?

****

LET THE VARIABLES BE CHOSEN AS SHOWN IN THE FIGURE.

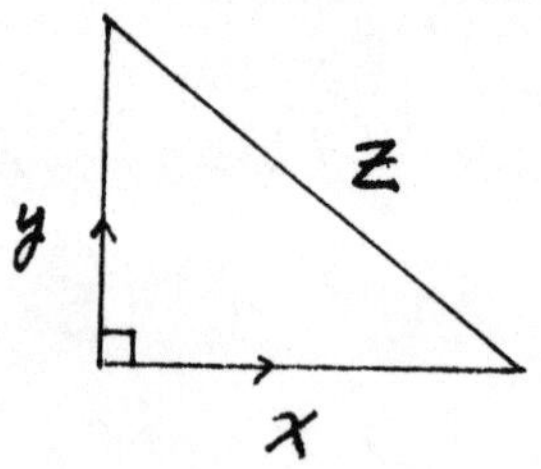

WE KNOW THAT

$$x^2 + y^2 = z^2$$

AND THAT

$$\frac{dx}{dt} = 5 \qquad \frac{dy}{dt} = 4.$$

WE WANT $\frac{dz}{dt}$ AT $t = 2$.

DIFFERENTIATING IMPLICITLY WITH RESPECT TO t,

$$2x\frac{dx}{dt} + 2y\frac{dy}{dt} = 2z\frac{dz}{dt}$$

$$x\frac{dx}{dt} + y\frac{dy}{dt} = z\frac{dz}{dt}.$$

AT $t = 2$, WE HAVE

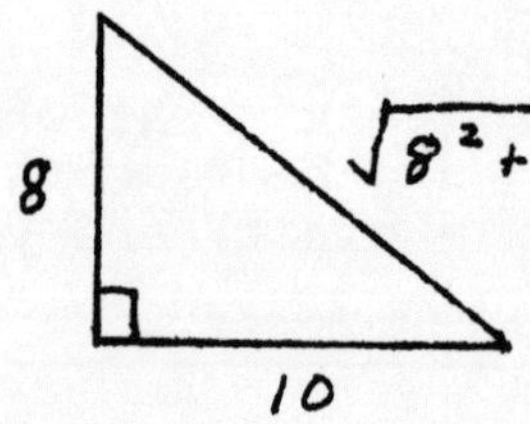

$$\sqrt{8^2 + 10^2} = \sqrt{164} = 2\sqrt{41}$$

SO

$$10 \cdot 5 + 8 \cdot 4 = 2\sqrt{41}\frac{dz}{dt}$$

$$82 = 2\sqrt{41}\frac{dz}{dt}$$

$$\frac{dz}{dt} = \frac{82}{2\sqrt{41}} = \frac{82\sqrt{41}}{2\sqrt{41}\sqrt{41}} = \sqrt{41}.$$

THE DISTANCE IS CHANGING AT $\sqrt{41}$ MI/HR.

4-52

A mothball shrinks in such a way that its radius decreases by 1/6th inch per month. How fast is the volume changing when the radius is 1/4th inch? Assume mothball is spherical.

****

"How fast" asks us to find the change in volume compared to the change in time; ie $\frac{dV}{dt}$

$\frac{dV}{dt} = \frac{dV}{dr} \cdot \frac{dr}{dt}$ Since $V = \frac{4}{3}\pi r^3$, $\frac{dV}{dr} = 4\pi r^2$. It's given that $\frac{dr}{dt} = \frac{-1}{6}$

So $\frac{dV}{dt} = 4\pi r^2 \cdot \frac{-1}{6} = \frac{-2\pi r^2}{3}$;

when $r = \frac{1}{4}$ this is $\frac{-\pi}{24}$ "/month

4-53

The electric resistance of a certain resistor as a function of temperature is given by $R = 6.000 + 0.002T^2$, where R is measured in Ohms and T in degrees Celsius. If the temperature is decreasing at the rate of 0.2°C per second, find the rate of change of resistance when T = 38°C.

****

GIVEN: $\frac{dT}{dt} = -0.2°C/\text{second}$

$$\left.\frac{dR}{dt}\right|_{T=38°} = \left.2(.002)T\frac{dT}{dt}\right|_{T=38°} = .004(38)(-.2)$$

$$= -0.0304 \text{ ohms/second}$$

4-54

A man 6 feet tall walks at the rate of 200 feet per minute towards a street light which is 18 feet above the ground. At what rate is the tip of his shadow moving?

****

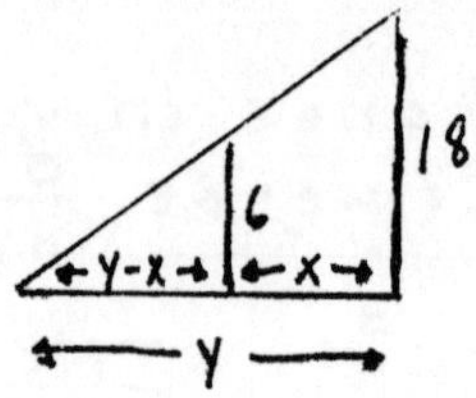

x = distance from man to light
y = distance from tip of shadow to light.
y − x = distance from tip of shadow to man.

USING SIMILAR TRIANGLES $\frac{18}{y} = \frac{6}{y-x}$

Then, $18y - 18x = 6y$

OR $12y = 18x$

OR $y = \frac{3}{2}x$

DIFFERENTIATING BOTH SIDE, WITH RESPEST TO TIME YIELDS $\frac{dy}{dt} = \frac{3}{2}\frac{dx}{dt}$

THEREFORE $\frac{dy}{dt} = \frac{3}{2}(200) = 300 \frac{\text{FT.}}{\text{MIN.}}$

4-55

A streetlight is 12 feet high. A moth is 2 feet from the lamppost and is flying straight up at 1 foot per second. How fast is its shadow moving along the ground when it's 11 feet off the ground?

****

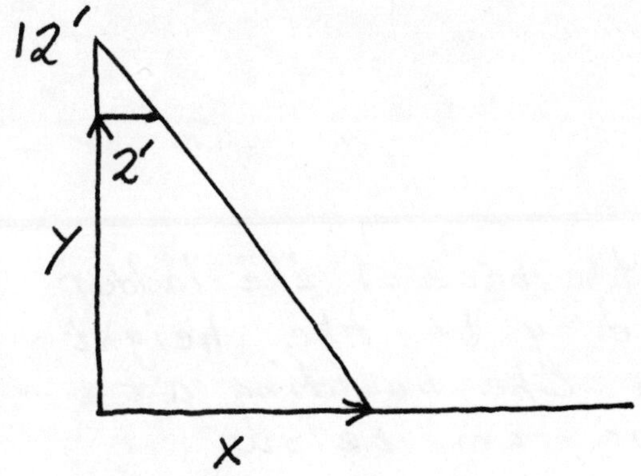

We want $\frac{dx}{dt}$ when $y = 11$. Using similar triangles, $\frac{x}{12} = \frac{2}{12-y}$, $x = \frac{24}{12-y}$.

Thus $\frac{dx}{dt} = 24 \frac{d}{dt}\left(\frac{1}{12-y}\right)$

$$= 24 \frac{(-1)(-1)}{(12-y)^2} \frac{dy}{dt}.$$

When $y = 11$, $\frac{dx}{dt} = 24$ ft/sec since $\frac{dy}{dt}$ is always 1 ft/sec.

4-56

Mr.Cooper is standing on the top of his 16 foot ladder when he realizes that the ladder is slipping down the side of the building. He decides that the base of the ladder is moving away from the bottom of the building at a rate of 2 feet per second when it is 3 feet from the bottom of the building. How fast is Mr. Cooper falling at that instant?

****

Let x be the distance of the base of the ladder from the building and y be the height of the ladder up on the building at time t. See the diagram below.

Given: $\frac{dx}{dt} = 2$ feet per second when $x = 3$.

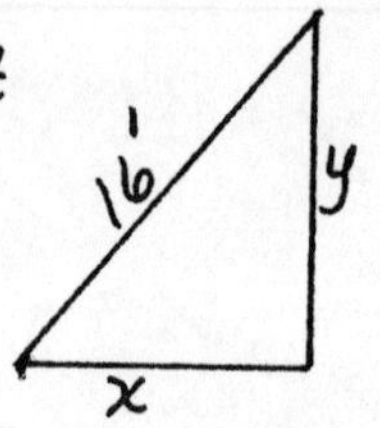

We want to find $\frac{dy}{dt}$ when $x = 3$ feet.

Since the ladder forms a rt. triangle, $x^2+y^2 = 256$. So if $x = 3$, $9+y^2=256$ and $y = \sqrt{247}$.

Differentiating $x^2+y^2=256$ with respect to t we get $x\frac{dx}{dt} + y\frac{dy}{dt} = 0$. Replace x by 3, $\frac{dx}{dt}$ by 2 and y by $\sqrt{247}$ to obtain

$\frac{dy}{dt} = -\frac{6}{\sqrt{247}}$ feet per second when $x = 3$.

Since the sign is negative, Mr. Cooper is <u>falling</u> at the rate of $\frac{6}{\sqrt{247}}$ feet per second.

CONCAVITY AND POINTS OF INFLECTION

4-57

Determine a so that the function $f(x) = x^2 + \frac{a}{x}$ has an inflection point at $x = 1$.

****

First, calculate $f''(x)$:

$$f(x) = x^2 + \frac{a}{x} = x^2 + ax^{-1}$$

$$f'(x) = 2x - ax^{-2}$$

$$f''(x) = 2 + 2ax^{-3} = 2\left(1 + \frac{a}{x^3}\right)$$

If $f(x)$ has an inflection point at $x = 1$, then $f''(1) = 0$

$$f''(1) = 2\left(1 + \frac{a}{1}\right) = 2(1 + a) = 0 \text{ when } \underline{a = -1}$$

Check: For $a = -1$, $f''(x) = 2\left(1 - \frac{1}{x^3}\right)$

$$\begin{cases} f''(x) < 0 \text{ when } \frac{1}{x^3} > 1 \text{ or } 0 < x < 1 \\ f''(x) > 0 \text{ when } x < 0 \text{ or } x > 1 \end{cases}$$

Hence $f(x)$ is concave down for $x < 1$

$f(x)$ is concave up for $x > 1$

Thus: $x = 1$ is an inflection point of $f(x) = x^2 + \frac{1}{x}$

4-58

Determine concavity for $f(x) = \frac{4}{x^2-1}$.

****

$$f'(x) = \frac{-4(2x)}{(x^2-1)^2} = \frac{-8x}{(x^2-1)^2}$$

$$f''(x) = \frac{(x^2-1)^2(-8) - (-8x)\cdot 2(x^2-1)(2x)}{(x^2-1)^4}$$

$$= \frac{(x^2-1)(-8) - (-8x)\cdot 2\cdot 2x}{(x^2-1)^3} = \frac{24x^2+8}{(x^2-1)^3}.$$

Now the graph of $f(x) = \frac{4}{x^2-1}$ is concave up if and only if $f''(x) > 0$. Since the numerator, $24x^2+8$, is always positive, $f''(x) > 0$ exactly when $x^2-1>0$. This is true when $x>1$ or when $x<-1$. The graph of $f(x)$ is concave up for all x less than -1, is concave down for all x between -1 and +1, and is concave up for all x greater than 1.

4-59

Find the values a,b,c so the function $f(x) = x^3 + ax^2 + bx + c$ has a critical point at (1,5) and an inflection point at (2,3).

****

$f'(x) = 3x^2 + 2ax + b \qquad f''(x) = 6x + 2a$

$f''(2) = 0 = 6(2) + 2a \rightarrow a = -6$

$f'(1) = 0 = 3(1)^2 + 2(-6)(1) + b \rightarrow b = 9$

$f(1) = 5 = (1)^3 + (-6)(1)^2 + 9(1) + c \rightarrow c = 1$

4-60

Given the function $f(x) = 2x^6 + 9x^5 + 10x^4 - 13x - 5$, determine all intervals on which the graph of f is concave up, all intervals where it is concave down, and find all inflection points for f.

****

$$f'(x) = 12x^5 + 45x^4 + 40x^3 - 13$$

$$f''(x) = 60x^4 + 180x^3 + 120x^2$$

$$= 60x^2(x^2 + 3x + 2)$$

$$= 60x^2(x+1)(x+2)$$

DERIVATIVE SIGN CHART

x		-2		-1		0	
f'	- - - irrelevant - - - - - -						
f''	+	0	−	0	+	0	+
f		21		11		-5	

$f''(1) > 0$

$f''(-1/2) > 0$

$f''(-3/2) < 0$

$f''(-3) > 0$

The graph of f(x) is:

concave up on the intervals $(-\infty, -2)$ and $(-1, \infty)$.

concave down on the interval $(-2, -1)$.

The points $(-2, 21)$ and $(-1, 11)$ are inflection points, while $(0, -5)$ is <u>not</u> an inflection point, since the second derivative does not change sign there.

NEWTON'S METHOD FOR ROOTS OF EQUATIONS

4-61

Use Newton's method to approximate a solution to the following equation.

$$x^3 + 2x = 3.1$$

****

Let $f(x) = x^3 + 2x - 3.1$

$f'(x) = 3x^2 + 2$

$f(x_{n+1}) - f(x_n) \approx f'(x_n)(x_{n+1} - x_n)$ and we want $f(x_{n+1})$ to be at least approximately 0.

Then $$x_{n+1} \approx \frac{f'(x_n)\,x_n - f(x_n)}{f'(x_n)} = \frac{2x_n^3 + 3.1}{3x_n^2 + 2}$$

$x_0 = 1$ by inspection of the given equation.

$x_1 = \frac{5.1}{5} = 1.02$ is an approximate solution.

4-62

Use Newton's algorithm to determine a root (to two decimal places) of the equation $x^3 + 8x - 23 = 0$, given an initial starting value of $x_o = 2$.

****

Newton's algorithm: $X_{n+1} = X_n - \dfrac{f(x_n)}{f'(x_n)}$

In this problem, we seek a zero of the function $f(x) = x^3 + 8x - 23$. Since $f'(x) = 3x^2 + 8$, then we have:

$$x_{n+1} = x_n - \frac{x_n^3 + 8x_n - 23}{3x_n^2 + 8}$$

$$x_0 = 2$$

$$x_1 = 2 - \frac{8 + 16 - 23}{12 + 8} = 2 - \frac{1}{20} = 1.95$$

$$x_2 = 1.95 - \frac{1.95^3 + 8(1.95) - 23}{3(1.95)^2 + 8} \approx 1.9492$$

$\therefore$ $x = 1.95$ is a root of the given equation, to two decimal places.

4-63

Use Newton's Method to find the root of $6x^3 + x^2 - 19x + 6 = 0$ that lies between 0 and 1.

****

Set $f(x) = 6x^3 + x^2 - 19x + 6$

$f'(x) = 18x^2 + 2x - 19$

$$x = x_0 - \frac{f(x_0)}{f'(x_0)} = x_0 - \frac{6x_0^3 + x_0^2 - 19x_0 + 6}{18x_0^2 + 2x_0 - 19}$$

Begin with $x_0 = 0$:

$$x = 0 - \frac{6}{-19} = \frac{6}{19} = 0.316$$

Now, let $x_0 = 0.316$:

$$x = 0.316 - \frac{6(.316)^3 + (.316)^2 - 19(.316) + 6}{18(.316)^2 + 2(.316) - 19}$$

$$= 0.316 - \frac{0.285}{-16.571} = 0.316 + 0.017 = 0.333$$

Let $x_0 = 0.333$:

$$x = 0.333 - \frac{6(.333)^3 + (.333)^2 - 19(.333) + 6}{18(.333)^2 + 2(.333) - 19}$$

$$= 0.333 - \frac{.0054}{-16.338} = 0.333 + 0.0003 = 0.333$$

The root is 0.333.

4-64

A rule for approximating cube roots states that $\sqrt[3]{a} \approx x_{n+1}$ where

$$x_{n+1} = \frac{1}{3}(2x_n + \frac{a}{x_n^2}) \qquad \text{for } n = 1, 2, 3, \ldots$$

and x_1 is any approximation to $\sqrt[3]{a}$.

a) Use Newton's Method to derive this rule.

b) Use the rule to approximate $\sqrt[3]{9}$.

****

a)

Consider the function: $f(x) = x^3 - a$

Note that $f(x) = 0 \Rightarrow x = \sqrt[3]{a}$

Now apply Newton's Method to approximate the zero of $f(x)$.

Since $f'(x) = 3x^2$ we have:

$$x_{n+1} = x_n - \frac{x_n^3 - a}{3x_n^2} = \frac{3x_n^3 - x_n^3 + a}{3x_n^2} = \frac{1}{3}\left(2x_n + \frac{a}{x_n^2}\right)$$

b)

To approximate $\sqrt[3]{9}$, choose $x_1 = 2$

then $x_2 = \frac{1}{3}\left(4 + \frac{9}{4}\right) \approx 2.0833333$

$x_3 \approx \underline{\underline{2.0800886}}$

Which is accurate to 5 decimal places.

GRAPH SKETCHING OF A FUNCTION

4-65

For the curve $f(x) = x^2/(x-2)^2$, find the domain and range, all asymptotes, intervals increasing, intervals concave up, local maximums, local minimums, inflection points, and sketch the curve.

****

$f(x) = \dfrac{x^2}{(x-2)^2}$ $\quad (x-2)^2 = 0$ at $x = 2$, vertical asymptote

$\lim_{x \to +\infty} f(x) = 1 \to y = 1$, horizontal asymptote

$f'(x) = \dfrac{-4x}{(x-2)^3}$ $\quad$ $-$ 0 $+$ 2 $-$ $\quad$ local minimum $(0,0)$

local maximum none

increasing $(0,2)$

$f''(x) = \dfrac{8(x+1)}{(x-2)^4}$ $\quad$ $-$ -1 $+$ 2 $+$ $\quad$ inflection point $(-1, \frac{1}{9})$

concave up $(-1,2)$ $(2,+\infty)$

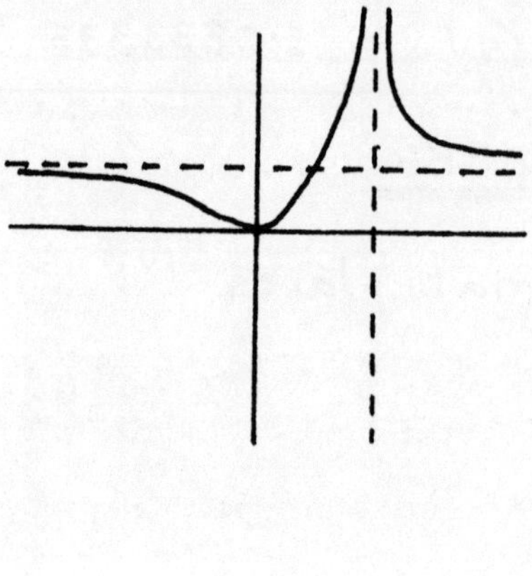

domain $\{x \mid x \neq 2\}$

range $\{y \mid y \geq 0\}$

4-66

Sketch a graph of y = (2x)/(x^2 + 1). Identify and label all asymptotes, extreme points and points of inflection.

****

asymptotes: vertical: none because denominator $\neq 0$ for any real values of x

horizontal: deg num < deg den. so x-axis is horiz. asymp.

oblique: none because deg num < deg den.

extreme values:

$$y' = \frac{(x^2+1)(2)-(2x)(2x)}{(x^2+1)^2} = \frac{2(1-x^2)}{(x^2+1)^2}$$

$$y'' = \frac{(x^2+1)^2(-4x)-(2-2x^2)(2)(x^2+1)(2x)}{(x^2+1)^4} = \frac{-4x(3-x^2)}{(x^2+1)^3}$$

$y'=0 \Rightarrow 1-x^2=0 \Rightarrow x=\pm 1$

by 2nd deriv test: $y''(1) = -1<0$ so $x=1$ is rel. max.

$y''(-1) = 1>0$ so $x=-1$ is rel min.

y' is defined everywhere and we have no domain endpoints so we have no other possible relative extremes

points of inflection: $y''=0 \Rightarrow x=0,\ 3-x^2=0 \Rightarrow x=\pm\sqrt{3}$

since y'' changes sign as it passes through these points we have three points of inflection

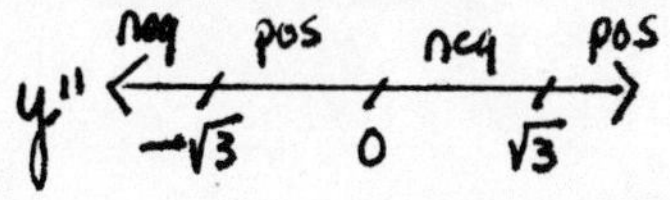

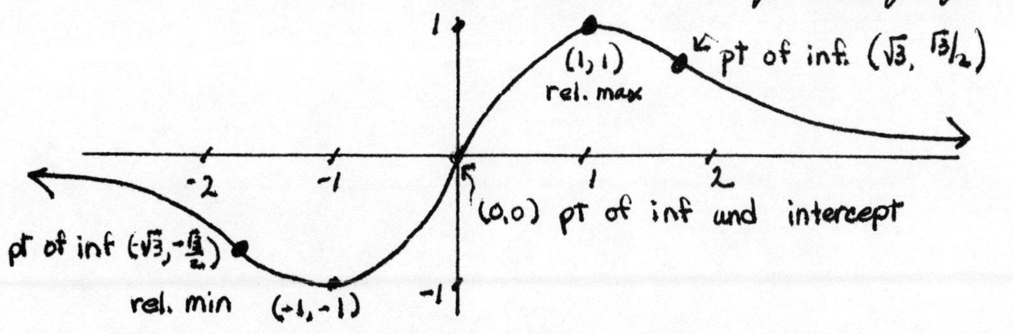

4-67

From the following information about the function f, sketch the graph of f:

Domain: all real numbers except -5.
The first derivative, f′, is such that:
$f'(x) > 0$ whenever $-5 < x < -3$ or $x > 0$;
$f'(x) < 0$ whenever $x < -5$ or $-3 < x < 0$;
$f'(x) = 0$ at the points (-3,1) and (0,-5).
The second derivative, f", is such that:
$f''(x) > 0$ whenever $-1 < x < 2$;
$f''(x) < 0$ whenever $x < -5$ or $-5 < x < -1$ or $x > 2$;
$f''(x) = 0$ at the points (-1,-3) and (2,-2).
x-intercepts: (-6,0), (-4,0) and (-2,0).
y-intercept: (0,-5).

$\lim_{x \to \infty} f(x) = 0$ $\lim_{x \to -\infty} f(x) = 2$ $\lim_{x \to -5} f(x) = -\infty$.

****

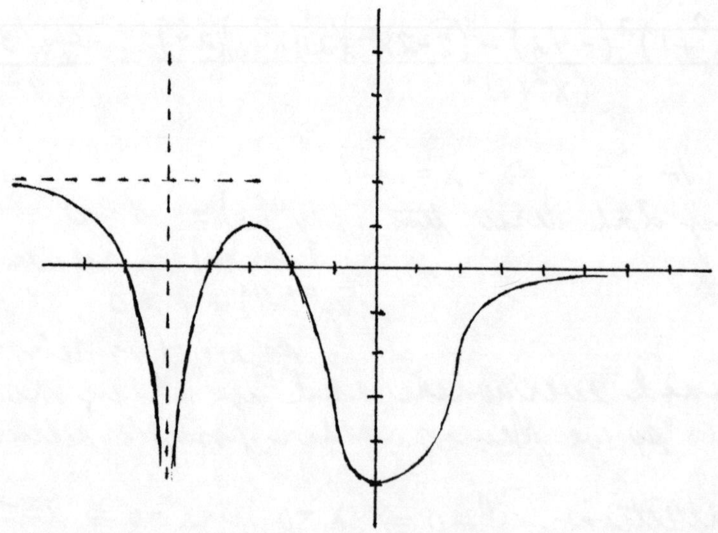

4-68

For the function $f(x) = x^3 - x^2 - 1$, $x \varepsilon [-2,3]$, (1) find all relative, endpoint, and absolute maxima and minima, (2) find all points of inflection, (3) find all horizontal and vertical asymptotes, (4) find where f is increasing and where it is decreasing, (5) find where f is concave upward and where it is concave downward, (6) sketch the graph of f.

****

$f'(x) = 3x^2 - 2x$

$f''(x) = 6x - 2$

$f'''(x) = 6$

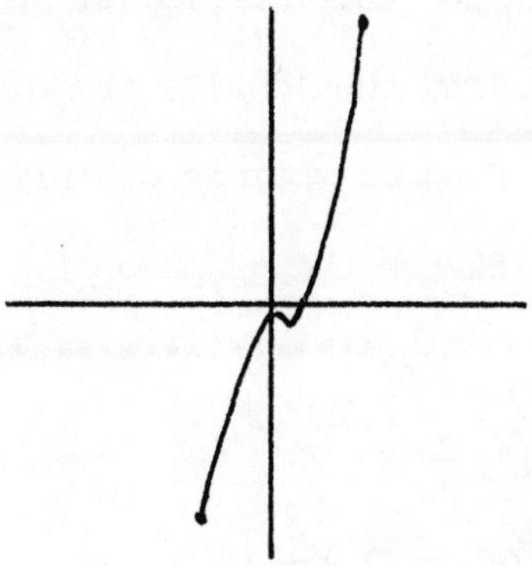

$3x^2 - 2x = 0$

$x = 0$ OR $x = \frac{2}{3}$

$f''(0) = -$

THUS, RELATIVE MAXIMUM OF -1 AT $x = 0$

$f''(\frac{2}{3}) = +$

THUS, RELATIVE MINIMUM OF $-\frac{31}{27}$ AT $x = \frac{2}{3}$

$6x - 2 = 0$

$x = \frac{1}{3}$

$f'''(\frac{1}{3}) = +$

THUS, POINT OF INFLECTION AT $(\frac{1}{3}, -\frac{29}{27})$

FROM THE GRAPH:

ENDPOINT MAXIMUM OF 17 AT $x = 3$

ENDPOINT MINIMUM OF -13 AT $x = -2$

ABSOLUTE MAXIMUM OF 17 AT $x = 3$

ABSOLUTE MINIMUM OF -13 AT $x = -2$

NO HORIZONTAL OR VERTICAL ASYMPTOTES

INCREASING ON $[-2,0] \cup [\frac{2}{3},3]$

DECREASING ON $[0,\frac{2}{3})$

CONCAVE UP ON $(\frac{1}{3},3]$

CONCAVE DOWN ON $[-2,\frac{1}{3})$

4-69

Consider the function $f(x) = \frac{1}{3}x^3 - \frac{1}{2}x^2 - 2x + 2$.

(a) Find the critical numbers.

(b) Find the interval(s) in which f is increasing.

(c) Find the interval(s) in which f is decreasing.

(d) Find the interval(s) in which the graph of f is concave upward.

(e) Find the interval(s) in which the graph of f is concave downward.

(f) Find the local extrema.

(g) Find any points of inflection.

(h) Sketch the graph of f.

****

$f'(x) = x^2 - x - 2$

$f''(x) = 2x - 1$

(a) SET $f'(x) = 0$. $\therefore (x-2)(x+1) = 0$
HENCE $x = -1, 2$ CRITICAL NUMBERS

(b) $f'(x) \geq 0$ FOR $(-\infty, -1] \cup [2, \infty)$
INCREASING INTERVALS

(c) $f'(x) \leq 0$ FOR $[-1, 2]$ DECREASING INTERVAL

(d) $f''(x) > 0$ FOR $(\frac{1}{2}, \infty)$ CONCAVE UPWARD

(e) $f''(x) < 0$ FOR $(-\infty, \frac{1}{2})$ CONCAVE DOWNWARD

(f) $f''(-1) < 0$ $\therefore$ LOCAL MAXIMUM AT $\left(-1, \frac{19}{6}\right)$.

$f''(2) > 0$ $\therefore$ LOCAL MINIMUM AT $\left(2, -\frac{4}{3}\right)$.

(g) POINT OF INFLECTION IS $\left(\frac{1}{2}, \frac{11}{12}\right)$.

(h)

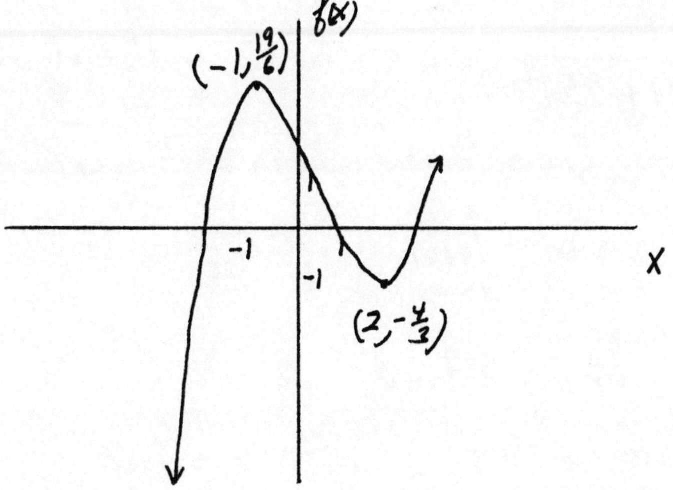

4-70

Sketch $y = \dfrac{x}{1 + x^2}$ showing all horizontal tangents and asymptotes.

****

$$y' = \frac{1(1+x^2) - x(2x)}{(1+x^2)^2} = \frac{1+x^2-2x^2}{(1+x^2)^2}$$

$$= \frac{1-x^2}{(1+x^2)^2}, \text{ so } y' = 0 \text{ when } x = \pm 1.$$

$y'(0) = 1, y(0) = 0, y(1) = \frac{1}{2}, y(-1) = -\frac{1}{2}$.

As $x \longrightarrow +\infty$, $\lim\limits_{x \to +\infty} \dfrac{x}{1+x^2} = \lim\limits_{x \to +\infty} \dfrac{1}{\frac{1}{x} + x}$

$= 0$. Thus there is a horizontal asymptote of 0 at $x = +\infty$.

$\lim\limits_{x \to -\infty} \dfrac{x}{1+x^2}$ also is 0, so there is a horizontal asymptote of 0 at $x = -\infty$.

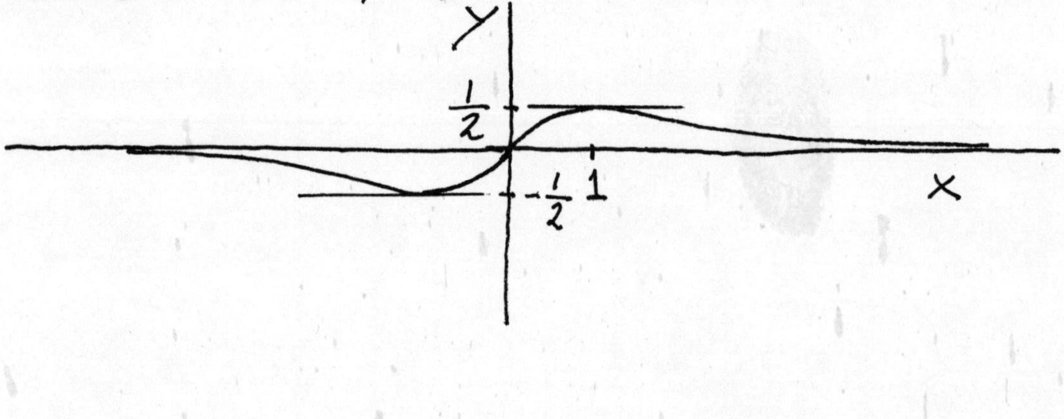

4-71

Mark the point (c,f(c)). Then sketch the part of the graph of y = f(x) near this point, given that f is continuous at c and the stated conditions:

a. $f'(c) = 0$, $f''(c) = -3$

b. $f'(c) = 1$ and $f''(c) = 2$

c. $f'(c) = -1$, $f''(x) < 0$ if $x < c$, $f''(x) > 0$ if $x > c$.

d. $f'(c)$ does not exist, $f'(x) > 0$ and $f''(x) > 0$ if $x < c$, $f'(x) < 0$ and $f''(x) > 0$ if $x > c$.

****

a.

(This is a local maximum point. It has a horizontal tangent and is concave downward.)

b.

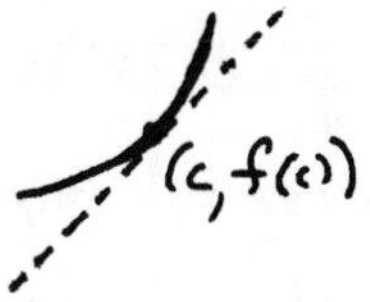

(It is helpful to draw a line with slope 1 first. Then draw a curve which is both tangent to the line and concave upward.)

c.

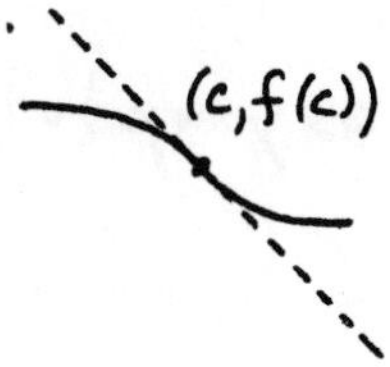

(First the line with slope -1. Since the curve is concave downward before, and concave upward after, it will be a point of inflection.)

d.

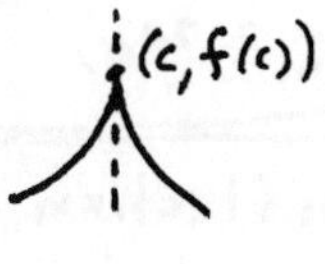

(Note that the curve increases and is concave upward before (c,f(c)) and decreases, but is concave upward, after it. Such a point is called a <u>cusp</u> on the graph.)

4-72

Sketch a complete and accurate graph of the curve $y = \frac{1}{x} - \frac{1}{x^2}$, including and labeling all extrema, inflection points, and asymptotes.

****

$$y' = -x^{-2} + 2x^{-3} = x^{-3}(2-x)$$

$$y'' = 2x^{-3} - 6x^{-4} = 2x^{-4}(x-3)$$

Derivative sign chart

x		0		2		3	
y'	↘	⋮	↗	0	↘		
y''	(−)	⋮		(−)		0	(+)
y		⋮		1/4		2/9	

$$\lim_{x \to \pm\infty} \frac{1}{x} - \frac{1}{x^2} = 0$$

$\Rightarrow$ $y = 0$ is a horizontal asymptote

$$y = \frac{x-1}{x^2} = 0 \text{ if } x = 1$$

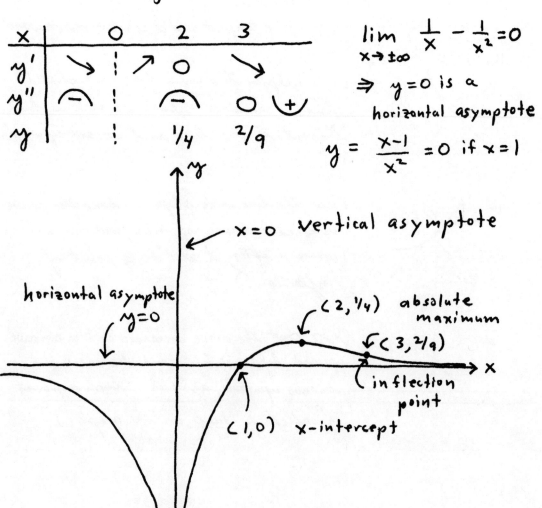

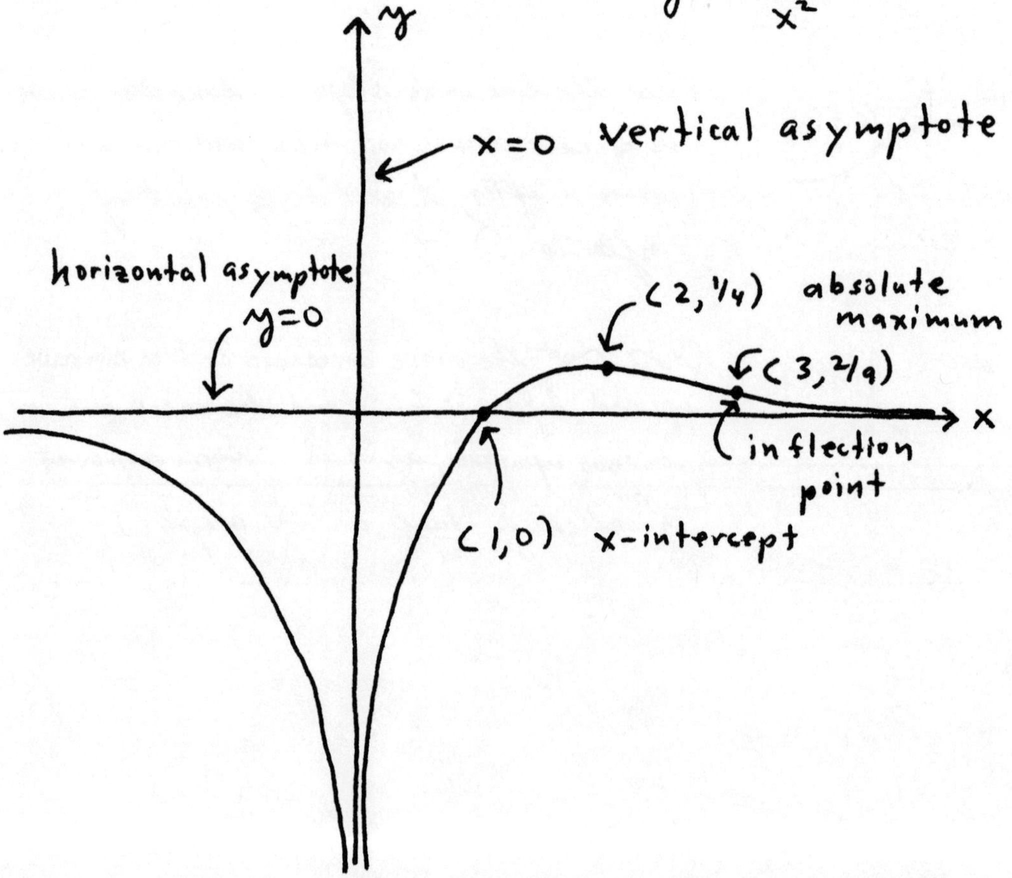

4-73

Given $f(x) = \frac{(x - 1)^2}{x^2 - 4}$ and $f'(x) = \frac{2(x - 1)(x - 4)}{(x^2 - 4)^2}$

sketch f(x)

****

From $f(x) = \frac{(x-1)^2}{x^2-4}$

X intercept, when $y=0$, $\rightarrow (x-1)^2=0 \rightarrow x=1$

Y intercept, when $x=0$, $\rightarrow y = \frac{(0-1)^2}{0-4} \rightarrow y = -\frac{1}{4}$

Vertical asymptotes when $x^2-4=0$, when $x=\pm 2$

Horizontal asymptote $\lim_{x \to \infty} \frac{(x-1)^2}{x^2-4} = 1$, at $y=1$

From $f'(x) = \frac{2(x-1)(x-4)}{(x^2-4)^2}$

	-2		1		2		4	
$2(x-1)$	- - -	- - -	+ + +	+ + +	+ + +
$(x-4)$	- - -	- - -	- - -	- - -	+ +
$(x^2-4)^2$	+ + +	+ + +	+ + +	+ + +	+ + +
$f'(x)$	+ ↗	VA	+ ↗	HT (1,0)	- ↘	VA	- ↘	HT $(4, \frac{9}{16})$	+ ↗

Horizontal Tangents at $(1,0)$, $(4, \frac{9}{16})$

Increasing $(-\infty,-2) \cup (-2,1] \cup [4,\infty)$

Decreasing $[1,2) \cup (2,4]$

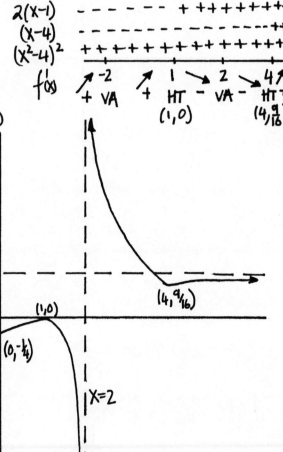

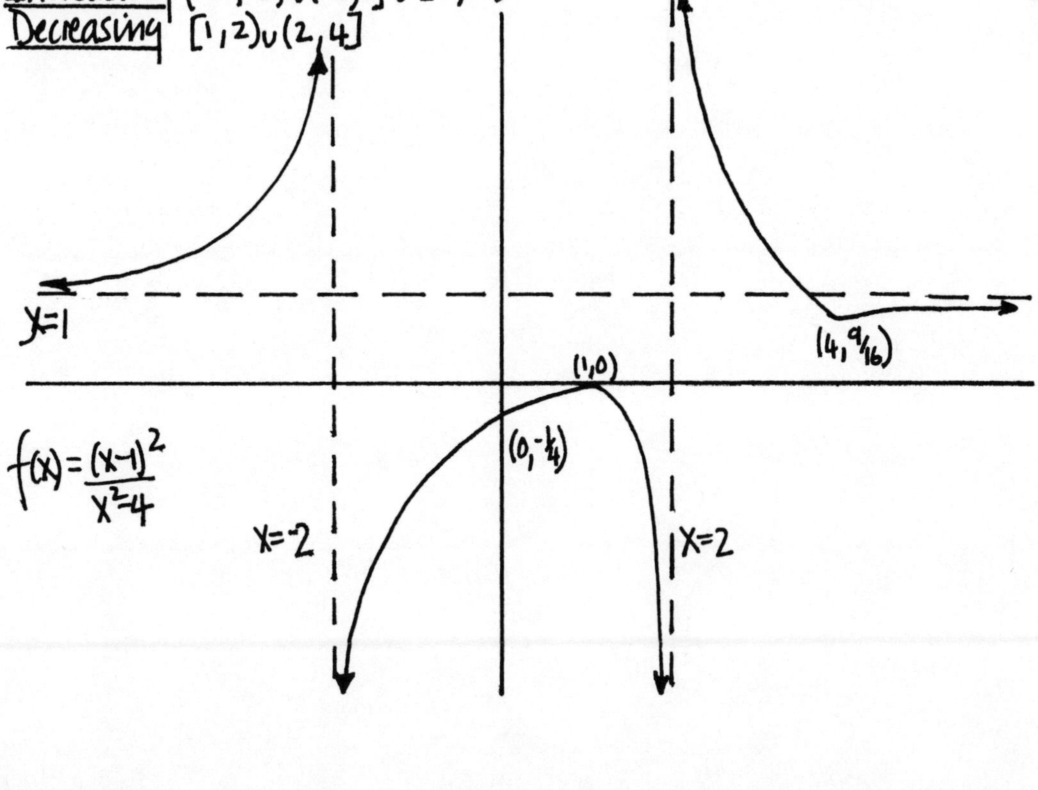

4-74

Sketch the graph of a function in a neighborhood of x = 2 which satisfies the following:

f(2) = 3
f'(2) = 2
f''(2) = -1

****

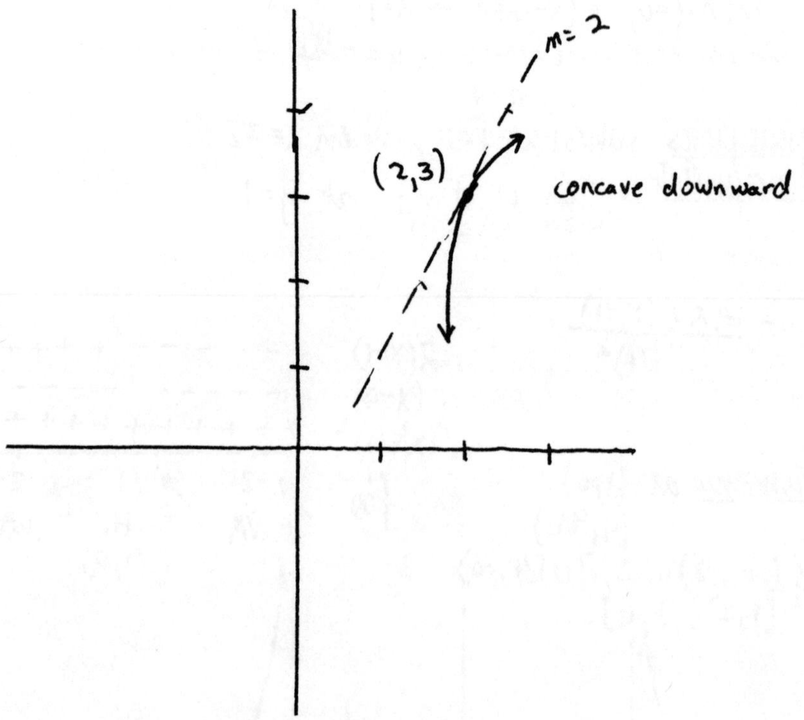

4-75

Let $f(x)=\frac{1}{3}x^3-\frac{1}{2}x^2$. Find the relative maximum, relative minimum, inflection points, the point where the graph of f crosses the x-axis, and the point where the graph crosses the y-axis. Then sketch the graph of f(x).

****

$f'(x)=x^2-x$. Let $f'(x)=0$; hence, $x^2-x=0$. So $x(x-1)=0$. So the critical values are $x=0$ and $x=1$. $f''(x)=2x-1$. Hence $f''(0)=-1<0$ and $(0,0)$ is a relative maximum. Also $f''(1)=2(1)-1>0$ so that $(1,-\frac{1}{6})$ is a relative minimum.

To find the inflection point, let $f''(x)=0$. So $2x-1=0$ and $x=\frac{1}{2}$ and the inflection point is $(\frac{1}{2},-\frac{1}{12})$.

Since $f(0)=0$ the graph of f crosses the y-axis at $(0,0)$ and since $(0,0)$ is a relative maximum, $x=0$ is a double root. Let $f(x)=0$. The roots of f are $x=0$ (a double root) and $x=\frac{3}{2}$. So f is tangent to the x-axis at $(0,0)$ and crosses the x-axis at $(\frac{3}{2},0)$. So the graph of f is given below:

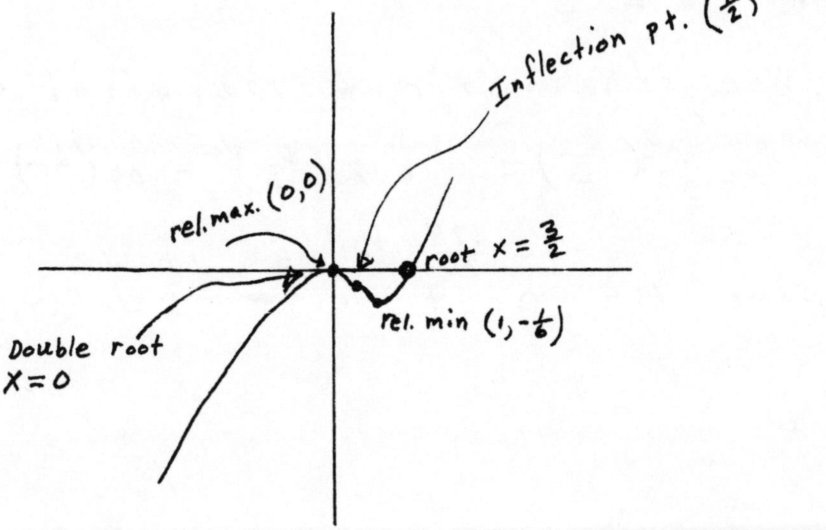

APPLICATIONS OF THE DERIVATIVE IN ECONOMICS

4-76

Suppose that the cost function for a certain article is given by $C(x) = .004x^3 - .02x^2 + 6x + 1000$. Find the minimum marginal cost.

****

By definition, the marginal cost m, is given by $m = C'(x)$. So

$$m = C'(x) = .012x^2 - .04x + 6$$

Hence $C''(x) = .024x - .04$

Let $C''(x) = 0$. $.024x - .04 = 0$

So

$$x = \frac{.04}{.024} = \frac{5}{3}$$

So the minimum marginal cost occurs when $x = \frac{5}{3}$.

Hence the minimum marginal cost is

$$C'\left(\frac{5}{3}\right) = .012\left(\frac{5}{3}\right)^2 - .04\left(\frac{5}{3}\right) + 6$$

$$= \frac{179}{30} \text{ dollars}$$

if we assume the cost unit is dollars.

4-77

If total cost, c, is related to sales, x, by $c = 0.6x^2 - 179x + 100$, find the amount of sales which will lead to maximum profit.

****

$$P = x - (0.6x^2 - 179x + 100)$$

$$P = -0.6x^2 + 180x - 100$$

$$\frac{dP}{dx} = -1.2x + 180$$

$$-1.2x + 180 = 0$$

$$x = \$150$$

$$\frac{d^2P}{dx^2} = -1.2$$

THUS, $x = 150$ LEADS TO MAXIMUM PROFIT

4-78

If the cost of manufacturing x units per day of a certain commodity is $C = 600 + .04x + .002x^2$ and if each unit sells for \$10.00, what daily production will maximize the profit?

****

PROFIT = REVENUE − COST i.e. $P = R - C$

$$P = 10x - (600 + .04x + .002x^2)$$

$$\frac{dP}{dx} = 10 - .04 - .002(2)x$$

SET $\frac{dP}{dx} = 0$ $\quad \therefore 0.004x = 9.96$

$$\therefore x = \frac{9.96}{.004} = 2490 \text{ UNITS}$$

4-79

Suppose that C(x) is the number of dollars in the total cost of producing x tables ($x \geq 6$) and

$$C(x) = 25 + 4x + \frac{18}{x}$$

Find: (a) the marginal cost function;

(b) the marginal cost when x = 10

****

a) Marginal cost is defined as the derivative of the cost function, therefore the answer is

$$C'(x) = 4 - \frac{18}{x^2}$$

b) $C'(10) = 4 - \frac{18}{(10)^2} = 4 - \frac{18}{100} = 3.82$ in dollars.

5
INTEGRALS

THE ANTIDERIVATIVE

5-1

(a) Find an antiderivative for $f(x) = \sin x$.

(b) Find $f(x)$ if $f'(x) = 3x - x^2$ and $f(1) = 4$.

****

(a)

$F(x) = -\cos x$, since

$$F'(x) = \frac{d}{dx}(-\cos x) = \sin x$$

(b)

The general antiderivative is $\frac{3}{2}x^2 - \frac{1}{3}x^3 + c$.

Since $f(1) = 4$, $4 = \frac{3}{2} \cdot 1^2 - \frac{1}{3} \cdot 1^3 + c = \frac{7}{6} + c$.

Thus $c = \frac{17}{6}$, and $f(x) = \frac{3}{2}x^2 - \frac{1}{3}x^3 + \frac{17}{6}$.

ANTIDIFFERENTIATION AND RECTILINEAR MOTION

5-2

An astronaut stands on a platform 3 meters above the moon's surface and throws a rock directly upward with an initial velocity of 32 meters/second.

a) Given that the acceleration due to gravity on the moon's surface is 1.6 meters/sec^2, then derive an equation which gives the height of the rock at time t seconds.

b) How high will the rock travel?

****

a) Since acceleration due to gravity produces negative velocity, we have

$$\text{Acceleration} = a(t) = -1.6 \text{ meters/sec}^2$$

Now, $\text{Velocity} = v(t) = \int a(t)\,dt = -1.6t + C$

But $v(0) = (-1.6)\,0 + C = 32$, so $C = 32$

Hence, $v(t) = -1.6t + 32$ meters/second

Also, $\text{Position} = s(t) = \int v(t)\,dt = \int(-1.6t + 32)\,dt$

$$= -0.8t^2 + 32t + C$$

But $s(0) = (-0.8)0^2 + 32\cdot 0 + C = 3$, so $C = 3$

Therefore $\underline{\underline{s(t) = -0.8t^2 + 32t + 3}}$ meters

b) Maximum height is attained when

$$v(t) = 0 \Rightarrow -1.6t + 32 = 0 \Rightarrow t = 20 \text{ sec.}$$

And $s(20) = (-0.8)20^2 + 32\cdot 20 + 3 = \underline{\underline{323}}$ meters.

5-3

For an object in rectilinear motion, suppose that the acceleration at time t is given by $a(t) = 12t - 6$. If the velocity at time t=2 is -3, and the position at time t=1 is 4, find the position function, s(t), for the object.

****

$$v(t) = \int a(t)\,dt = \int (12t-6)\,dt = 6t^2 - 6t + c$$

$$-3 = v(2) = 6(4) - 6(2) + c = 12 + c, \text{ so } c = -15$$

$$s(t) = \int v(t)\,dt = \int (6t^2 - 6t - 15)\,dt = 2t^3 - 3t^2 - 15t + k$$

$$4 = s(1) = 2 - 3 - 15 + k = k - 16, \text{ so } k = 20$$

$$\therefore \quad s(t) = 2t^3 - 3t^2 - 15t + 20$$

5-4

Find the position function s(t) given acceleration $a(t) = 3t$ if $v(2) = 0$ and $s(2) = 1$.

****

$$\int a(t) = v(t) = \int 3t\,dt = \frac{3t^2}{2} + C$$

$$v(2) = 0 \Rightarrow 0 = \frac{3(2)^2}{2} + C \Rightarrow C = -6$$

$$\text{so } v(t) = \frac{3t^2}{2} - 6$$

$$s(t) = \int v(t) = \int \left(\frac{3t^2}{2} - 6\right)dt = \frac{t^3}{2} - 6t + C$$

$$s(2) = 1 \Rightarrow 1 = \frac{2^3}{2} - 6(2) + C \Rightarrow C = 9$$

$$\text{so } s(t) = \frac{t^3}{2} - 6t + 9$$

5-5

A particle moving along the number line has acceleration given by $a(t) = 2t-1$

$V_0 = -2$ (V_0 = velocity at time $t = 0$)

$S_0 = +2$ (S_0 = position at time $t = 0$)

a) What is the <u>net</u> distance traversed from time $t = 0$ to $t = 3$?

b) What is the <u>total</u> distance traversed from time $t = 0$ to $t = 3$?

****

First, find the velocity $v(t)$:

$$v(t) = \int a(t)\,dt = \int (2t-1)\,dt = t^2 - t + C$$

Since $v_0 = -2$, $C = -2$ and $v(t) = t^2 - t - 2$

a) for the <u>net</u> distance $S = \int_0^3 v(t)\,dt$

$$= \int_0^3 (t^2 - t - 2)\,dt = \frac{t^3}{3} - \frac{t^2}{2} - 2t \Big]_0^3 = \frac{27}{3} - \frac{9}{2} - 2\cdot 3$$

$$= -3/2$$

b) for the <u>total</u> distance $S = \int_0^3 |v(t)|\,dt$

We need to find when $v(t) > 0$:

$v(t) = t^2 - t - 2 = (t-2)(t+1) > 0$ when $t < -1$ or $t > 2$

Hence:

$$\int_0^3 |v(t)|\,dt = -\int_0^2 (t^2 - t - 2) + \int_2^3 (t^2 - t - 2)\,dt$$

$$= -\left(\frac{t^3}{3} - \frac{t^2}{2} - 2t\right]_0^2 + \left(\frac{t^3}{3} - \frac{t^2}{2} - 2t\right]_2^3 = \frac{10}{3} + \left(-\frac{3}{2} - \left(-\frac{10}{3}\right)\right)$$

$= \frac{20}{3} - \frac{3}{2} = \frac{31}{6} = 5\frac{1}{6}.$

Note: the information $S_0 = +2$ is irrelevant to the solution of the problems.

5-6

The acceleration of a moving particle starting at a position 100 m along the x-axis and moving with an initial velocity of $v_0 = 25$ m/min is given by $a(t) = 13\sqrt{t}$ m/min^2. Find the equation of motion of the particle.

****

$$v(t) = \int 13\sqrt{t}\,dt$$
$$= 13\left(\frac{2}{3}\right)t^{3/2} + C$$
$$= \frac{26}{3}t^{3/2} + C$$
$$v(0) = \frac{26}{3}(0)^{3/2} + C = 25$$
$$C = 25$$
$$v(t) = \frac{26}{3}t^{3/2} + 25$$

$$x(t) = \int\left(\frac{26}{3}t^{3/2} + 25\right)dt$$
$$= \frac{26}{3}\left(\frac{2}{5}\right)t^{5/2} + 25t + K$$
$$= \frac{52}{15}t^{5/2} + 25t + K$$
$$x(0) = K = 100$$
$$K = 100$$
$$x(t) = \frac{52}{15}t^{5/2} + 25t + 100$$

5-7

A bag of diamonds is accidentally dropped from a building 98m high. How long does it take for the bag to hit the ground? (Assume that g=9.8 meters per second per second).

****

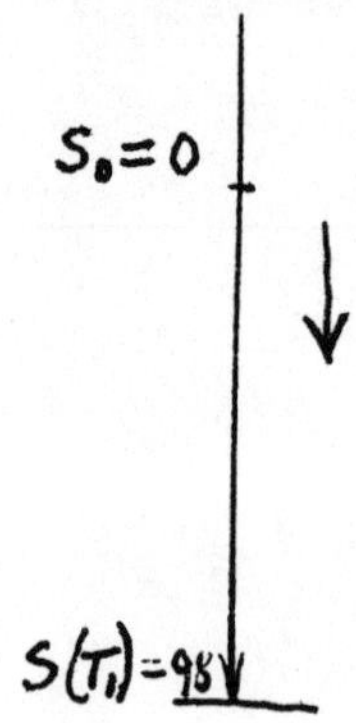

Note: Since the bag is dropped by accident, there is no initial velocity imparted to it, so $v_o=0$;

Let us assume that the initial position is $S_o=0$, as shown. Then at the time when the bag hits the ground, say at $t=t_1$, the position function is $S(t_1)=98$.

SINCE THE ONLY FORCE ACTING ON THE FALLING BAG IS THE FORCE OF GRAVITY,

$$MA = Mg \rightarrow a = g \quad (1)$$

SINCE $a = \frac{dV}{dt}$ AND $g = 9.8\ ^{m}/_{sec^2}$

EQUATION (1) BECOMES: $\frac{dV}{dt} = 9.8$ SOLVING THIS SIMPLE DIFFERENTIAL EQUATION FOR $V(t)$, WE HAVE:

$$V(t) = 9.8t + V_0 \rightarrow V(t) = 9.8t \quad (2) \quad (V_0 = 0)$$

SINCE $V(t) = \frac{ds}{dt}$, EQUATION (2) BECOMES: $\frac{ds}{dt} = 9.8t$

SOLVING THIS EQUATION FOR $S(t)$, WE HAVE:

$$s(t) = 4.9t^2 + S_0 \rightarrow s(t) = 4.9t^2 \quad (3) \quad (s_0 = 0)$$

FINALLY, AT $t = t_1$, WHEN THE BAG HITS THE GROUND, $s(t_1) = 98\ m$ THUS (3) BECOMES:

$$98 = 4.9t_1^2 \rightarrow t_1 = \sqrt{\frac{98}{4.9}} = \sqrt{20} = 2\sqrt{5} \doteq 4.5\ \text{SEC.}$$

<u>ALTERNATIVE APPROACH</u>

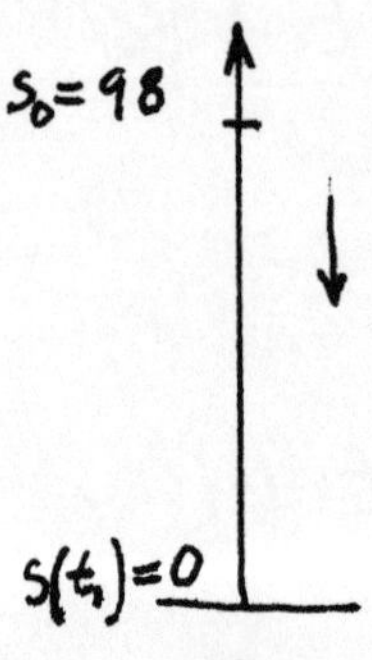

NOTE: $V_0=0$ FOR THE REASONS STATED ABOVE. WE ASSUME THE INITIAL POSITION OF THE BAG TO BE $S_0 = 98$ AS SHOWN. THEN AT THE TIME WHEN THE BAG HITS THE GROUND, SAY $t = t_1$, THE POSITION FUNCTION IS $s(t_1) = 0$

SINCE THE ONLY FORCE ACTING ON THE BAG IS THE FORCE OF GRAVITY, WE HAVE $MA = -Mg \rightarrow a = -g$ (1)
(THE MINUS SIGN IS CHOSEN SINCE WE ASSUME UPWARD MOVEMENT TO BE POSITIVE).

FROM (1) WE HAVE:

$$\frac{dV}{dt} = -9.8 \quad \left(\text{SINCE } a = \frac{dV}{dt} \text{ AND } g = 9.8\right)$$

SOLVING THIS EQUATION FOR $V(t)$, WE GET:

$$V(t) = -9.8t + V_0 \rightarrow V(t) = -9.8t \quad (2) \quad (V_0 = 0)$$

FROM (2) WE HAVE: $\frac{ds}{dt} = -9.8t$ $\left(\text{SINCE } V = \frac{ds}{dt}\right)$

SOLVING IT FOR $s(t)$, WE OBTAIN:

$$s(t) = -4.9t^2 + S_0 \rightarrow s(t) = -4.9t^2 + 98 \quad (\text{SINCE } S_0 = 98)$$

FINALLY, AT $t = t_1$ (WHEN THE BAG HITS THE GROUND) $s(t_1) = 0$, AND WE HAVE:

$$-4.9t^2 + 98 = 0 \rightarrow t_1 = \sqrt{\frac{98}{4.9}} = \sqrt{20} = 2\sqrt{5} \doteq 4.5 \text{ SEC}$$

THE FUNDAMENTAL THEOREM OF CALCULUS

5-8

Find $\frac{dy}{dx}$ & $\frac{d^2y}{dx^2}$ if $y = \int_1^{3x} \frac{dt}{t^2 + t + 1}$

****

Let $u = 3x$

$$\Rightarrow y = \int_1^{u} \frac{dt}{t^2+t+1}$$

$$\Rightarrow \frac{dy}{du} = \frac{1}{u^2+u+1}$$ (By the Fund. Thm. of Calc.)

But, using the chain rule,

$$\frac{dy}{dx} = \frac{dy}{du} \cdot \frac{du}{dx}$$

$$\Rightarrow \frac{dy}{dx} = \left(\frac{1}{u^2+u+1}\right)(3)$$

$$\Rightarrow \frac{dy}{dx} = \frac{3}{9x^2+3x+1}$$ where $3x$ is substituted for u

$$\Rightarrow \frac{d^2y}{dx^2} = \frac{-3(18x+3)}{(9x^2+3x+1)^2}$$

5-9

Let $f(x) = \int_0^x \frac{t^2 - 4}{1 + \cos^2 t}\,dt$. Find and classify the relative maximums and minimums of $f(x)$.

****

By the Fundamental theorem of Calculus

$$f'(x) = \frac{x^2 - 4}{1 + \cos^2 x}$$

Note: $f'(x)$ exists for all x. To find the extreme values of $f(x)$, set $f'(x) = 0$

$$f'(x) = \frac{x^2 - 4}{1 + \cos^2 x} = 0 \quad \text{when} \quad x = \pm 2$$

Note that the denominator of $f'(x)$ is always positive

Hence $\begin{cases} f'(x) > 0 & \text{for } x^2 - 4 > 0 \text{ i.e. } |x| > 2 \\ f'(x) < 0 & \text{for } |x| < 2 \end{cases}$

Hence $f(x)$ $\begin{cases} \text{increasing for } |x| > 2 \\ \text{decreasing for } |x| < 2 \end{cases}$

therefore has $f(x)$ has rel. max at $x = -2$

and has rel. min at $x = +2$

5-10

Find y' if $y = \int_1^x e^{(t^2)}dt$.

****

By the Fundamental Theorem of Calculus,

$$y' = e^{(x^2)}.$$

5-11

$$\frac{d}{dt}\left(\int_{t^2}^{2} \sqrt{x+1}\, dx\right) =$$

****

RECALL.

$$\frac{d}{dt}\int_{f(t)}^{g(t)} F(x)\, dx = F(g(t))g'(t) - F(f(t))f'(t).$$

HERE $g(t) = 2$ so $g'(t) = 0$

$f(t) = t^2$ so $f'(t) = 2t$

and $F(x) = \sqrt{x+1}$

So

$$\frac{d}{dt}\int_{t^2}^{2} \sqrt{x+1}\, dx = \left(-\sqrt{t^2+1}\right) 2t.$$

THE DEFINITE INTEGRAL

5-12

Evaluate $\int_{-3}^{4} \left| |x| - 4 \right| dx$.

****

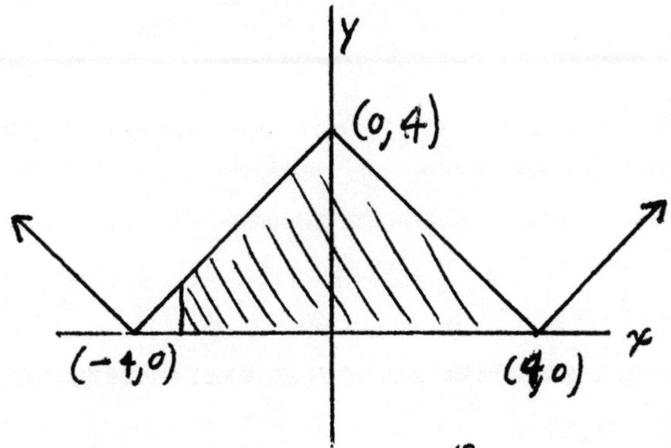

$$\int_{-3}^{4} \left| |x| - 4 \right| dx = \int_{-3}^{0} (x+4)\,dx + \int_{0}^{4} (4-x)\,dx$$

$$= \frac{x^2}{2} + 4x \Big|_{-3}^{0} + 4x - \frac{x^2}{2} \Big|_{0}^{4}$$

$$= (0+0) - \left(\frac{9}{2} - 12\right) + \left(16 - \frac{16}{2}\right) - (0-0)$$

$$= -\frac{9}{2} + 12 + 16 - 8 = 20 - \frac{9}{2}$$

$$= \frac{40-9}{2} = \frac{31}{2}.$$

5-13

Solve for x when $\int_0^1 t^x\,dt = 5$.

****

Since t is the variable of integration we may consider x a constant.

$$\int_0^1 t^x\,dt = \left.\frac{t^{x+1}}{x+1}\right|_0^1 = \frac{1}{x+1} - \frac{0}{x+1} = \frac{1}{x+1} = 5.$$

$$\frac{1}{5} = x+1 \Rightarrow x = -4/5.$$

5-14

Given:

$$\int_0^3 f(x)\,dx = 12 \qquad \text{and} \qquad \int_0^6 f(x)\,dx = 42$$

Assume that f is continuous everywhere and use the properties of the definite integral to evaluate the following

$$\int_3^6 [2f(x) - 3]\,dx$$

****

$$\int_3^6 [2f(x) - 3]\,dx = 2\int_3^6 f(x) - 3\int_3^6 dx$$

$$= 2\left[\int_0^6 f(x)\,dx - \int_0^3 f(x)\,dx\right] - 3\int_3^6 dx$$

$$= 2[42 - 12] - 3x\Big|_3^6$$

$$= 2(30) - 3(6-3) = 51$$

5-15

If $\int_0^5 kx\,dx = 30$, find k.

****

("k" is treated just like any constant.)

$$\int_0^5 kx\,dx = \frac{kx^2}{2}\Big|_0^5 = \frac{25k}{2}$$

Since this equals 30, we have

$$\frac{25k}{2} = 30$$

$$25k = 60$$

$$k = \frac{12}{5}$$

5-16

Find $\int_{-1}^{0} \frac{4p^2-8p+1}{(p-1)^2}\,dp$

****

$$\int_{-1}^{0} \frac{4p^2-8p+1}{(p-1)^2}\,dp = \int_{-1}^{0} \frac{4p^2-8p+1}{p^2-2p+1}\,dp = \int_{-1}^{0}\left(4 - \frac{3}{p^2-2p+1}\right)dp$$

$$= \int_{-1}^{0}\left[4 - 3(p-1)^{-2}\right]dp = 4p - \frac{3(p-1)^{-1}}{-1}\Big|_{-1}^{0}$$

$$= 4p + \frac{3}{p-1}\Big|_{-1}^{0} = -3 - \left(-4 - \frac{3}{2}\right) = \frac{5}{2}.$$

5-17

Find the average value of $f(x) = x^2 + 6x - 5$ on $[1,4]$.

****

$$\text{Average Value} = \frac{\int_1^4 (x^2+6x-5)\,dx}{4-1}$$

$$= \frac{\frac{1}{3}x^3 + 3x^2 - 5x \Big|_1^4}{3} = \frac{1}{3}\left[\left(\frac{64}{3}+48-20\right)-\left(\frac{1}{3}+3-5\right)\right]$$

$$= \frac{1}{3}\left[49\tfrac{1}{3} - \left(-1\tfrac{2}{3}\right)\right] = \frac{1}{3}(51) = 17$$

INTEGRATION OF POWERS OF SINE AND COSINE

5-18

What is the value of $\int_{\frac{\pi}{2}}^{\pi} (\sin 2x)\,dx$?

****

$$\int_{\frac{\pi}{2}}^{\pi} (\sin 2x)\,dx = \frac{1}{2}(-\cos 2x)\Big]_{\frac{\pi}{2}}^{\pi} =$$

$$-\frac{1}{2}\cos 2\pi - \left(-\frac{1}{2}\cos 2\cdot\frac{\pi}{2}\right) = -\frac{1}{2}(1) - \left(-\frac{1}{2}\right)(-1) = -1$$

5-19

Evaluate the following indefinite integral:

$$\int \sin^3(2x)\cos^4(2x)\,dx$$

****

$$\int \sin^3(2x)\cos^4(2x)\,dx = \int \sin^2(2x)\cos^4(2x)\cdot\sin(2x)\,dx$$

$$= \int [1-\cos^2(2x)]\cos^4(2x)\sin(2x)\,dx$$

$$= \int [\cos^4(2x) - \cos^6(2x)]\sin(2x)\,dx$$

$$= \int \cos^4(2x)\sin(2x)\,dx - \int \cos^6(2x)\sin(2x)\,dx$$

Since $d[\cos(2x) = -\sin(2x)\cdot 2\,dx$,

$$\int \sin^3(2x)\cos^4(2x)\,dx = -\tfrac{1}{2}\int \cos^4(2x)\,d[\cos(2x)] + \tfrac{1}{2}\int \cos^6(2x)\,d[\cos(2x)]$$

$$= -\tfrac{1}{2}\,\frac{\cos^5(2x)}{5} + \tfrac{1}{2}\,\frac{\cos^7(2x)}{7} + C$$

$$= -\tfrac{1}{10}\cos^5(2x) + \tfrac{1}{14}\cos^7(2x) + C$$

5-20

Evaluate $\int \sin^2 3x dx$.

****

Using a trig. identity,

$$\int \sin^2 3x\,dx = \int \frac{1-\cos 6x}{2}\,dx$$

$$= \int \frac{1}{2}\,dx - \frac{1}{2}\int \cos 6x\,dx$$

$$= \frac{1}{2}x - \frac{1}{12}\sin 6x + C.$$

5-21

$\int \sin 2x \cos^3 2x\, dx =$

****

$$\int \sin 2x \cos^3 2x\,dx = -\frac{1}{2}\int u^3\,du = -\frac{1}{2}\,\frac{u^4}{4} + C$$

$u = \cos 2x$

$du = -2\sin 2x\,dx$

SO SUBSTITUTE BACK TO GET

$$\int \sin 2x \cos^3 2x\,dx = -\frac{1}{8}\cos^4 2x + C.$$

5-22

Evaluate the following integral:

$$\int \sin^3 x \cos^3 x \, dx$$

****

Method 1

$$\int \sin^3 x \cos^3 x \, dx$$

$$= \int \sin^3 x \cos^2 x \cos x \, dx$$

$$= \int \sin^3 x (1 - \sin^2 x) \cos x \, dx$$

$$= \int (\sin^3 x - \sin^5 x) \cos x \, dx \qquad \text{let } u = \sin x, \; du = \cos x \, dx$$

the integral becomes

$$\int (u^3 - u^5) \, du$$

$$= \frac{u^4}{4} + -\frac{u^6}{6} + C$$

$$= \boxed{\frac{\sin^4 x}{4} - \frac{\sin^6 x}{6} + C}$$

Method 2 involves changing $\sin^2 x$ to $1 - \cos^2 x$

we evaluate

$$\int (\cos^3 x - \cos^5 x) \sin x \, dx, \qquad \text{letting } u = \cos x, \; du = -\sin x \, dx$$

the integral becomes

$$\int (u^5 - u^3) \, du = \boxed{\frac{\cos^6 x}{6} - \frac{\cos^4 x}{4} + C}$$

Note that the two methods give results that differ by a constant.

5-23

Evaluate the following $\int \sin^2 x \cos^4 x \, dx$

****

$$\int \sin^2 x \cos^4 x \, dx = \int \left(\frac{1-\cos 2x}{2}\right)\left(\frac{1+\cos 2x}{2}\right)^2 dx \quad \text{(half-angle formulas)}$$

$$= \frac{1}{8}\int (1-\cos 2x)(1+\cos 2x)^2 dx = \frac{1}{8}\int (1+\cos 2x - \cos^2 2x - \cos^3 2x)\, dx$$

$$= \frac{1}{8}\int \left(1 + \cos 2x - \left(\frac{1+\cos 4x}{2}\right) - \cos 2x(1-\sin^2 2x)\right) dx$$

$$= \frac{1}{8}\int \left(\frac{1}{2} - \frac{1}{2}\cos 4x + \cos 2x \sin^2 2x\right) dx$$

$$= \int \frac{1}{16}\, dx - \int \frac{1}{16}\cos 4x\, dx + \int \frac{1}{8}\sin^2 2x \cos 2x\, dx$$

$$\int \frac{1}{16}\, dx = \frac{1}{16}x \,, \quad \int \frac{1}{16}\cos 4x\, dx = \frac{1}{16}\,\frac{\sin 4x}{4} = \frac{1}{64}\sin 4x$$

$$\int \frac{1}{8}\sin^2 2x \cos 2x\, dx \quad (u = \sin 2x,\ du = 2\cos 2x\, dx)$$

$$= \int \frac{1}{8} u^2 \frac{1}{2}\, du = \frac{1}{16}\cdot\frac{u^3}{3} = \frac{\sin^3 2x}{48}$$

$$\therefore \int \sin^2 x \cos^4 x\, dx = \frac{1}{16}x - \frac{1}{64}\sin 4x + \frac{1}{48}\sin^3 2x + C$$

INTEGRATION BY SUBSTITUTION

5-24

Evaluate $\int_0^1 \frac{x^3 dx}{\sqrt{1+x^2}}$

****

this integral yields to the substitution:

$$u = 1+x^2 \qquad du = 2x\,dx$$

$$x^2 = u-1 \qquad x\,dx = \frac{1}{2}du \qquad x^3 dx = \frac{1}{2}(u-1)du$$

Note also that when $x=0$, $u=1$

when $x=1$, $u=2$

Hence, $\int_0^1 \frac{x^3 dx}{\sqrt{1+x^2}} = \frac{1}{2}\int_1^2 \frac{u-1}{\sqrt{u}}du$

$$= \frac{1}{2}\int_1^2 (u^{1/2} - u^{-1/2})du = \frac{1}{2}\left(\frac{2}{3}u^{3/2} - 2u^{1/2}\right]_1^2$$

$$= \frac{1}{2}\left[\left(\frac{2}{3}\cdot 2^{3/2} - 2\cdot 2^{1/2}\right) - \left(\frac{2}{3}-2\right)\right]$$

$$= \frac{1}{2}\left[-\frac{1}{3}\cdot 2^{3/2} + \frac{4}{3}\right] = \frac{1}{6}\left(4 - 2\sqrt{2}\right)$$

5-25

Evaluate $\int t\sqrt{1 + 5t^2}\,dt$.

****

Let $u = 1+5t^2$, $du = 10t\,dt$.

$$\int t\sqrt{1+5t^2}\,dt = \frac{1}{10}\int \sqrt{u}\,du$$

$$= \frac{1}{10}\cdot\frac{2}{3}u^{3/2} + c$$

and since $u = 1+5t^2$, our integral is $\frac{1}{15}(1+5t^2)^{3/2} + c$.

5-26

Evaluate the following: $\int x^2\sqrt[4]{x^3+2}\,dx$

****

Let $u = x^3+2$, then $du = 3x^2dx$, or $x^2dx = \frac{1}{3}du$

$$\int x^2\sqrt[4]{x^3+2}\,dx = \int\left(\sqrt[4]{x^3+2}\right)x^2dx = \int\sqrt[4]{u}\,\frac{du}{3}$$

$$= \frac{1}{3}\int u^{1/4}du = \frac{1}{3}\,\frac{u^{5/4}}{5/4} + C = \frac{1}{3}\cdot\frac{4}{5}u^{5/4} + C$$

$$= \frac{4}{15}(x^3+2)^{5/4} + C$$

5-27

Find (a) $\int \frac{1}{\sqrt{x}(1 + \sqrt{x})^2} dx$ (b) $\int x^3 \sqrt{x^2 + 4}\, dx$

****

(a) $\int \frac{1}{\sqrt{x}(1+\sqrt{x})^2} dx$

let $u = 1 + x^{\frac{1}{2}}$ then $\frac{du}{dx} = \frac{1}{2}x^{-\frac{1}{2}} = \frac{1}{2\sqrt{x}}$

solve for dx by cross multiplication

$2\sqrt{x}\, du = dx.$

$= \int \frac{1}{\cancel{\sqrt{x}}(u)^2} \cdot 2\cancel{\sqrt{x}}\, du$

$= \int 2u^{-2} du = \frac{2u^{-1}}{-1} + C = \frac{-2}{u} + C$

$= \underline{\frac{-2}{1+\sqrt{x}} + C}$

(b) $\int x^3 \sqrt{x^2+4}\, dx$

follow the same process used in part (a)

let $u = x^2 + 4$ then $\frac{du}{dx} = 2x$

solve for dx, $du = 2x\,dx$ or $dx = \frac{du}{2x}$

$= \int x^3 \sqrt{u} \cdot \frac{du}{2x}$

$= \int \frac{x^2}{2} \sqrt{u}\, du$

we need to replace x^2 with u's.
we know $u = x^2 + 4$ so $x^2 = u - 4$

$= \int \left(\frac{u-4}{2}\right) u^{\frac{1}{2}} du = \frac{1}{2}\int \left(u^{3/2} - 4u^{1/2}\right) du = \frac{u^{5/2}}{5} + \frac{4u^{3/2}}{3} + C = \underline{\frac{(x^2+4)^{5/2}}{5} + \frac{4(x^2+4)^{3/2}}{3} + C}$

5-28

One of the following definite integrals can be evaluated using the fundamental theorem of calculus while the other can not. Tell which one can and evaluate it.

a) $\int_0^1 \sqrt{x}\ \sqrt[3]{1 - x}\ dx$

b) $\int_0^1 \sqrt{1 - x}\ \sqrt[3]{1 - x}\ dx$

****

Problem b) is the one which can be evaluated.

let $u = 1 - x$

$du = -dx$

then $\int_0^1 \sqrt{1-x}\ \sqrt[3]{1-x}\ dx$ becomes

$$\int_{u(0)}^{u(1)} \sqrt{u}\ \sqrt[3]{u}\ (-du) = -\int_1^0 u^{1/2} u^{1/3}\ du$$

$$= \int_0^1 u^{5/6}\ du = \frac{6}{11} u^{11/6} \Big|_0^1$$

$$\boxed{= 6/11}$$

5-29

Evaluate: $\int_1^3 \frac{x\,dx}{\sqrt{1 + 3x^2}}$

****

let $u = 1 + 3x^2$

$du = 6x\,dx \Rightarrow \frac{1}{6}du = x\,dx$

Also $x = 1 \Rightarrow u = 1 + 3(1)^2 = 4$

$x = 3 \Rightarrow u = 1 + 3(3)^2 = 28$

Substituting:

$$\int_1^3 \frac{x\,dx}{\sqrt{1+3x^2}} = \int_4^{28} \frac{\frac{1}{6}du}{u^{1/2}} = \frac{1}{6}\int_4^{28} u^{-1/2}\,du = \frac{1}{6}\,\frac{u^{1/2}}{(1/2)}\Big]_4^{28}$$

$$= \frac{1}{3}\left(28^{1/2} - 4^{1/2}\right) = \frac{2}{3}\left(\sqrt{7} - 1\right)$$

5-30

Evaluate the following definite integral:

$$\int_1^9 \frac{3x}{\sqrt{10-x}}\,dx$$

****

Use the substitution $u = 10 - x$, $du = -dx$

$x = 10 - u$

$$\int_1^9 \frac{3x}{\sqrt{10-x}}\,dx = \int_9^1 \frac{-3(10-u)}{\sqrt{u}}\,du$$

$$= \int_1^9 (30u^{-1/2} - 3u^{1/2})\,du = 60u^{1/2} - 2u^{3/2}\Big|_1^9$$

$$= 180 - 54 - (60-2)$$

$$= 126 - 58 = 68$$

5-31

Evaluate $\displaystyle\int_1^8 \frac{4(x^{2/3}+14)^3}{\sqrt[3]{x}}\,dx$

****

$$\int_1^8 \frac{4(x^{2/3}+14)^3}{\sqrt[3]{x}}\,dx = 6\int_1^8 \frac{2}{3}x^{-1/3}(x^{2/3}+14)^3\,dx$$

$$= 6\left[\frac{(x^{2/3}+14)^4}{4}\right]_1^8$$

$$= \frac{3}{2}\left[(18)^4-(15)^4\right]$$

$$= \frac{163053}{2}$$

5-32

Evaluate the integral: $\int x^2\sqrt{1+x^3}\,dx.$

****

LET $u = 1+x^3$

THEN $du = 3x^2\,dx$

$$\int x^2\sqrt{1+x^3}\,dx = \frac{1}{3}\int (1+x^3)^{1/2}\,3x^2\,dx$$

$$= \frac{1}{3}\int u^{1/2}\,du$$

$$= \frac{1}{3}\,\frac{u^{3/2}}{3/2} + C$$

$$= \frac{2}{9}(1+x^3)^{3/2} + C$$

5-33

Find the following indefinite integral

$$\int x^2 \sqrt{2x - 1}\ dx$$

****

Let $u = 2x - 1$. Then $du = 2\,dx$ and $x = \frac{u+1}{2}$.

By substitution $\int x^2 \sqrt{2x-1}\,dx = \int \left(\frac{u+1}{2}\right)^2 \cdot u^{1/2} \cdot \frac{du}{2}$

$$= \int \frac{1}{8}(u^2 + 2u + 1) u^{1/2}\, du$$

$$= \tfrac{1}{8} \int (u^{5/2} + 2u^{3/2} + u^{1/2})\, du$$

$$= \tfrac{1}{8} \left[\tfrac{2}{7} u^{7/2} + \tfrac{4}{5} u^{5/2} + \tfrac{2}{3} u^{3/2}\right] + C$$

$$= \tfrac{1}{28}(2x-1)^{7/2} + \tfrac{1}{10}(2x-1)^{5/2} + \tfrac{1}{12}(2x-1)^{3/2} + C$$

5-34

Find $\int (2x^3+3x^2)^{2/3}(x^2+x)dx$.

****

If we let $u(x) = 2x^3 + 3x^2$, then $du = 6x^2 + 6x$. Thus the integrand will have the form $u^n du$ if we insert a factor of 6 inside the integral. Of course, we must then compensate by inserting a factor of 1/6 outside the integral sign.

Consequently, $\int (2x^3+3x^2)^{2/3}(x^2+x)\,dx$

$$= \frac{1}{6}\int (2x^3+3x^2)^{2/3}(6x^2+6x)\,dx$$

$$= \frac{1}{6}\cdot\frac{(2x^3+3x^2)^{5/3}}{5/3} + C = \frac{(2x^3+3x^2)^{5/3}}{10} + C.$$

5-35

Evaluate the following integral: $\int \frac{rdr}{(r^2+b^2)^{\frac{3}{2}}}$ where b is a constant.

****

Let $u = r^2 + b^2$ $\qquad du = 2r\,dr$

$\frac{1}{2}du = rdr$

$$\int \frac{rdr}{(r^2+b^2)^{\frac{3}{2}}} = \int \frac{\frac{1}{2}du}{u^{\frac{3}{2}}} = \frac{1}{2}\int u^{-\frac{3}{2}}du$$

$$= \frac{1}{2}\frac{u^{-\frac{1}{2}}}{-\frac{1}{2}} + C = -\frac{1}{u^{\frac{1}{2}}} + C$$

$$= -\frac{1}{(r^2+b^2)^{\frac{1}{2}}} + C = -\frac{1}{\sqrt{r^2+b^2}} + C$$

Thus, $\int \frac{rdr}{(r^2+b^2)^{\frac{3}{2}}} = -\frac{1}{\sqrt{r^2+b^2}} + C.$

5-36

Evaluate $\int_0^4 x\sqrt{x^2 + 9}\ dx$

****

Use integration by substitution, letting

$$u = x^2 + 9, \quad du = 2x\,dx.$$

Multiply and divide by 2 to convert the integral to an $\int u^{1/2}\,du$ form.

$$\int_0^4 x\sqrt{x^2+9}\,dx = \frac{1}{2}\int_0^4 (x^2+9)^{1/2}(2x)\,dx$$

This is of the form $\frac{1}{2}\int u^{1/2}\,du$, and an antiderivative is $\left(\frac{1}{2}\right)\frac{u^{3/2}}{3/2}$. So

$$\frac{1}{2}\int_0^4 (x^2+9)^{1/2}(2x)\,dx = \frac{1}{2}\left.\frac{(x^2+9)^{3/2}}{3/2}\right|_0^4$$

$$= \frac{1}{3}(x^2+9)^{3/2}\Big|_0^4 = \frac{1}{3}\left[(4^2+9)^{3/2} - (0^2+9)^{3/2}\right]$$

$$= \frac{1}{3}\left(25^{3/2} - 9^{3/2}\right) = \frac{1}{3}(125-27) = \frac{98}{3}$$

5-37

Find the anti-derivative: $\displaystyle\int \frac{4x - 4}{\sqrt{3x^2 - 6x}}\,dx$

****

Using u-substitution, choose $u = 3x^2 - 6x$

$$du = (6x-6)\,dx$$

$$dx = \frac{du}{6x-6}$$

Substituting:

$$\int \frac{4(x-1)}{u^{1/2}} \cdot \frac{du}{6(x-1)} =$$

$$\frac{2}{3}\int \frac{du}{u^{1/2}} = \frac{2}{3}\left(2u^{\frac{1}{2}}\right)$$

Replacing u with $3x^2 - 6x$, the answer is

$$\frac{4}{3}(3x^2 - 6x)^{1/2} + C$$

Checking:

$$\frac{d'\left(\frac{4}{3}(3x^2-6x)^{\frac{1}{2}} + C\right)}{dx} = \frac{4}{3} \cdot \frac{1}{2}(3x^2-6x)^{-\frac{1}{2}}(6x-6)$$

$$\frac{4}{6} \cdot \frac{1}{(3x^2-6x)^{1/2}} \cdot 6(x-1) = \frac{4x-4}{\sqrt{3x^2-6x}}$$

DIFFERENTIAL EQUATIONS WITH VARIABLES SEPARABLE

5-38

Find a particular solution to the separable differential equation

$$x^2y' = y^3, \quad y(1) = 1$$

****

$$x^2y' = y^3 \Rightarrow x^2\frac{dy}{dx} = y^3 \Rightarrow x^2dy = y^3dx$$

$$\Rightarrow \frac{dy}{y^3} = \frac{dx}{x^2}$$

so $\int y^{-3}dy = \int x^{-2}dx$

$$\frac{y^{-2}}{-2} = \frac{x^{-1}}{-1} + C \Rightarrow \frac{-1}{2y^2} = -\frac{1}{x} + C$$

substitute $x=1$ and $y=1$ to find C:

$$\frac{-1}{2(1^2)} = -\frac{1}{1} + C \Rightarrow C = 1/2$$

so $\frac{-1}{2y^2} = \frac{-1}{x} + \frac{1}{2}$

$$\frac{1}{2y^2} = \frac{2-x}{2x} \Rightarrow 2y^2 = \frac{2x}{2-x} \Rightarrow y^2 = \frac{x}{2-x}$$

5-39

Obtain a general solution of the following differential equation

$$\csc(x)\, dy + y^4\, dx = 0$$

****

Rewrite the equation as

$$\csc(x)\, dy = -y^4\, dx.$$

Separating variables,

$$\frac{dy}{y^4} = -\frac{dx}{\csc(x)}$$

or

$$y^{-4}\, dy = -\sin(x)\, dx.$$

Integrating gives the general solution.

$$\frac{y^{-3}}{-3} = \cos(x) + C$$

or

$$0 = \frac{1}{3y^3} + \cos(x) + C$$

5-40

Solve the Differential Equation $\frac{dz}{dt} = 2z^2t\sqrt{1+t^2}$, if z = 1 when t = 0.

****

SEPARATE VARIABLES

$$\frac{dz}{2z^2} = t\sqrt{1+t^2}\,dt.$$

$$\frac{1}{2}\int z^{-2}dz = \int t\sqrt{1+t^2}\,dt.$$

if $u = 1+t^2$

$du = 2t\,dt.$

$$-\frac{1}{2}z^{-1} = \frac{1}{2}\int u^{1/2}du.$$

$$-\frac{1}{2z} = \frac{1}{2}\left[\frac{2}{3}u^{3/2} + C\right]$$

SUBSTITUBE BACK TO GET. (CANCEL ½)

$$-\frac{1}{z} = \frac{2}{3}(1+t^2)^{3/2} + C.$$

TO FIND C; LET z = 1 AND t = 0

$$-1 = \frac{2}{3}(1) + C.$$

$$-5/3 = C$$

SO

$$-1/z = \frac{2}{3}(1+t^2)^{3/2} - 5/3.$$

5-41

Find the general solution of the following differential equation

$$\frac{dy}{dx} = \frac{(3x + 1)^2 \sqrt{2 - y^2}}{y}$$

****

Separating variables

$$\frac{y\,dy}{\sqrt{2-y^2}} = (3x+1)^2 dx$$

or $$(2-y^2)^{-1/2} y\,dy = (3x+1)^2 dx$$

Integrating

$$-\tfrac{1}{2}\,(2-y^2)^{1/2} \cdot 2 = \tfrac{1}{3}\,\frac{(3x+1)^3}{3} + C$$

$$-\frac{1}{\sqrt{2-y^2}} = \tfrac{1}{9}\,(3x+1)^3 + C$$

or $$0 = \tfrac{1}{9}(3x+1)^3 + \frac{1}{\sqrt{2-y^2}} + C$$

APPLICATIONS OF INTEGRATION IN ECONOMICS

5-42

An investment in a new piece of machinery will save a company money at the rate of $1600 per month initially, but the savings decline over time according to the following equation

$$s(t) = 1600 - \tfrac{1}{4}t^2$$

where t is time measured in months and s(t) is savings per month in dollars. Find the total savings accumulated by the new piece of machinery.

****

The total savings accumulated is the shaded area.

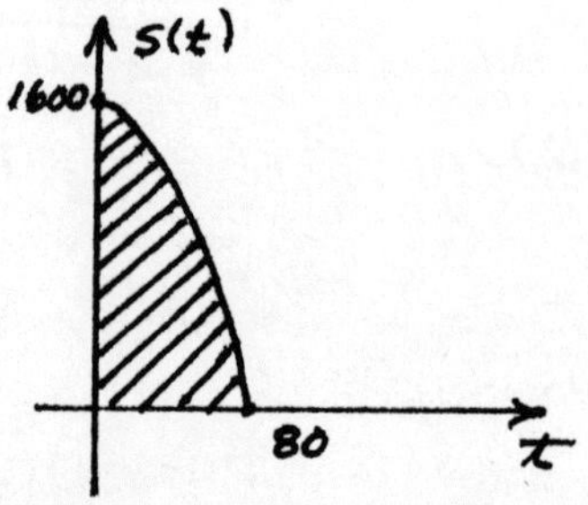

$$\therefore \text{Total Savings} = \int_0^{80} s(t)\,dt$$

$$= \int_0^{80} (1600 - \tfrac{1}{4}t^2)\,dt$$

$$= 1600t - \tfrac{1}{2}t^3 \Big|_0^{80}$$

$$= \$85{,}333.33$$

5-43

If the price of a new car is changing at a rate of 100 + $150\sqrt{t}$ dollars per year, find how much a new car will cost in 4 years if its present price is $8,000.

****

Call the rate of change of the price of the car $p'(t)$

Thus $p'(t) = 100 + 150\sqrt{t}$

Since we want the price in 4 years we need to find the function $p(t)$

$$p = \int p'(t)\,dt = \int (100 + 150\sqrt{t})\,dt = 100t + \frac{150t^{3/2}}{3/2} + c$$

so $p(t) = 100t + 100t^{3/2} + c$

To find the constant c, use the information that the initial price of the car was \$8,000

That is $p(0) = 8000$

$$p(0) = 100(0) + 100(0) + c = 8000 \rightarrow c = 8000.$$

hence $p(t) = 100t + 100t^{3/2} + 8000$

So the price after 4 years is given by $p(4)$

$$p(4) = 400 + 100(8) + 8000 = \$9{,}200$$

The price of a car in 4 years will be \$9,200

5-44

A company's marginal profit is given by:

$$P'(x) = \frac{2x}{\sqrt{2x^2 - 400}}$$

Use antidifferentiation to determine P(x) if P(100) = 0.

****

The family of profit functions is found by taking the antiderivative of marginal profit.

So $P(x) = \int \frac{2x}{\sqrt{2x^2 - 400}}\, dx$ Let $u = 2x^2 - 400$

$\frac{du}{dx} = 4x \Rightarrow dx = \frac{du}{4x}$

$$= \int \frac{2x}{\sqrt{u}} \frac{du}{4x}$$

$$= \int \frac{u^{-1/2}}{2}\, du = u^{1/2} + C = (2x^2 - 400)^{1/2} + C$$

Now, $P(100) = (19600)^{1/2} + C = 140 + C$

But $P(100) = 0$, so $C = -140$.

Hence $\underline{\underline{P(x) = (2x^2 - 400)^{1/2} - 140}}$

RIEMANN SUM APPROXIMATIONS

5-45

Find upper and lower bounds for $\int_1^6 \frac{dx}{x}$ by calculating the upper and lower Riemann sums, using the partition $x_0 = 1$, $x_1 = 3$, $x_2 = 6$ of the interval [1,6].

****

Sketch:

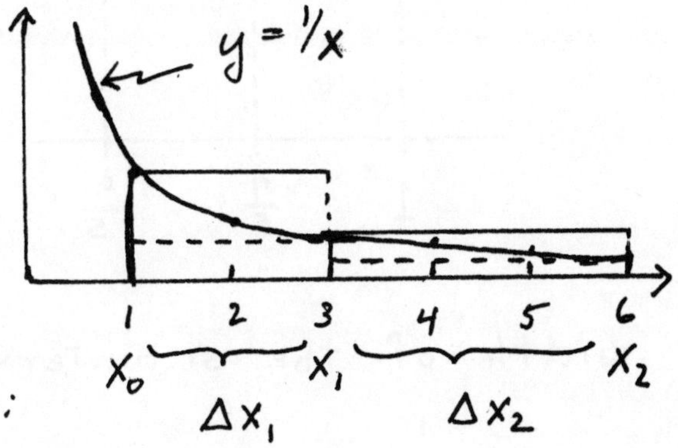

the Upper Riemann Sum (URS) is represented by the two rectangles with the solid horizontal roofs:

$$URS = f(x_0)\,\Delta x_1 + f(x_1)\,\Delta x_2$$

$$= 1\cdot 2 + \frac{1}{3}\cdot 3 = 2+1 = 3$$

the Lower Riemann Sum (LRS) is represented by the two rectangles with the dashed rectangular roofs:

$$LRS = f(x_1)\,\Delta x_1 + f(x_2)\,\Delta x_2$$

$$= \frac{1}{3}\cdot 2 + \frac{1}{6}\cdot 3 = \frac{2}{3} + \frac{1}{2} = \frac{7}{6} = 1\tfrac{1}{6}.$$

Hence: $1\tfrac{1}{6} < \int_1^6 \frac{dx}{x} < 3$

5-46

Use the Riemann sum corresponding to 5 inscribed rectangles of equal width to approximate the integral $\int_1^3 \frac{1}{x}\,dx$.

****

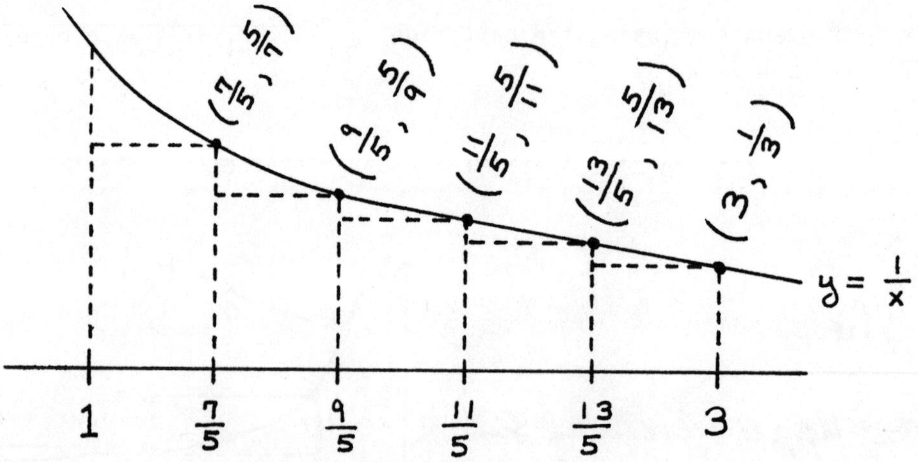

$$\text{width of the subintervals} = \frac{3-1}{5} = \frac{2}{5}$$

$$\text{Riemann sum} = \frac{2}{5}\left(\frac{5}{7} + \frac{5}{9} + \frac{5}{11} + \frac{5}{13} + \frac{1}{3}\right)$$

6 APPLICATIONS OF INTEGRALS

AREA OF A REGION IN A PLANE

6-1

Find the area bounded by the curves $f(x) = x^3 + x^2$ and $g(x) = 2x^2 + 2x$.

****

$x^3 + x^2 = 2x^2 + 2x \rightarrow x^3 - x^2 - 2x = 0 \rightarrow x(x+1)(x-2) = 0$

points of intersection $(0,0)$, $(-1,0)$, $(2,12)$

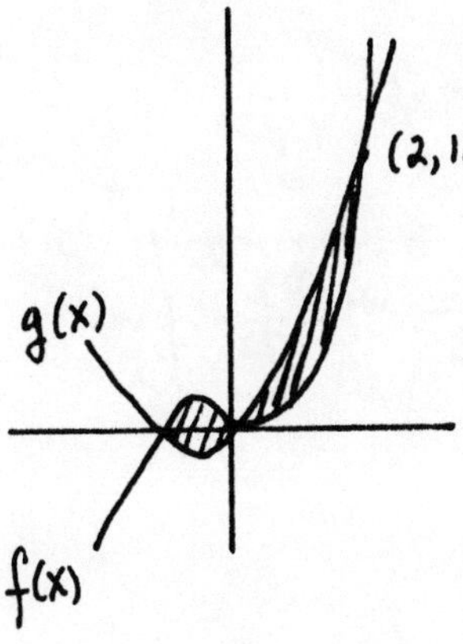

$$A = \int_{-1}^{0} [f(x) - g(x)]dx + \int_{0}^{2} [g(x) - f(x)]dx$$

$$= \int_{-1}^{0} (x^3 - x^2 - 2x)\,dx + \int_{0}^{2} (-x^3 + x^2 + 2x)\,dx$$

$$= \left(\frac{x^4}{4} - \frac{x^3}{3} - x^2\right)\Bigg|_{-1}^{0} + \left(-\frac{x^4}{4} + \frac{x^3}{3} + x^2\right)\Bigg|_{0}^{2}$$

$$= \left[(0) - \left(\frac{-5}{12}\right)\right] + \left[\left(\frac{8}{3}\right) - (0)\right] = \frac{37}{12} \text{ units}^2$$

6-2

Find the area of the region bounded by the graphs of f and g where

$$f(x) = x^2 - 4x + 5 \quad \text{and} \quad g(x) = 5 - x.$$

****

Set $f(x) = g(x)$ and solve.

$$x^2 - 4x + 5 = 5 - x$$

$$x^2 - 3x = 0$$

$$x(x-3) = 0$$

$x = 0$ or $x = 3$. Since f and g intersect where $x = 0$ and $x = 3$, determine which is larger over the interval 0 to 3. For example pick $x = 1$.

$$f(1) = 2 \text{ and } g(1) = 4.$$

$\therefore\ g(x) \geq f(x)$ when $0 \leq x \leq 3$.

$$\text{Area} = \int_0^3 [g(x) - f(x)]\,dx$$

$$\text{Area} = \int_0^3 [(5-x) - (x^2 - 4x + 5)\,dx$$

$$\text{Area} = \int_0^3 (-x^2 + 3x)\,dx = -\frac{x^3}{3} + \frac{3x^2}{2}\Big|_0^3$$

Area $= 9/2$ sq. units.

6-3

Given A: $y = x^2 - 1$ and B: $y = -x^2 + x + 2$.
Sketch the region bounded by the curves A and B, labelling all points of intersection. Find the area of the bounded region - be sure to show an appropriate representative rectangle.

****

A: parabola, opens up $(a = 1 > 0)$
vertex: $x = -b/2a = 0/2 = 0$
$y = 0^2 - 1 = -1$
$(0, -1)$

B: parabola, opens down
vertex: $x = -b/2a$
$= -1/2(-1) = 1/2$
$y = -(1/2)^2 + 1/2 + 2$
$= 9/4$
$(1/2, 9/4)$

pts. of intersection

$x^2 - 1 = -x^2 + x + 2$
$2x^2 - x - 3 = 0$
$(2x - 3)(x + 1) = 0$
$x = 3/2,\ x = -1$
$y = 9/4,\ y = 0$

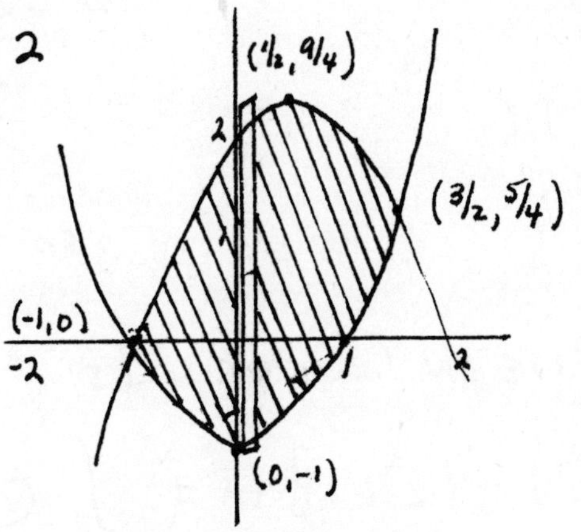

area of rep. rect. = (length)(width)
$= [(-x^2 + x + 2) - (x^2 - 1)]\,dx$

total area $= \int_{-1}^{3/2} [(-x^2 + x + 2) - (x^2 - 1)]\,dx$

$= \int_{-1}^{3/2} (-2x^2 + x + 3)\,dx$

$= \left(-\frac{2}{3}x^3 + \frac{1}{2}x^2 + 3x\right)\Big|_{-1}^{3/2} = 125/24$ square units

6-4

Find the area between the curves $y = 4 - x^2$ and $y = x^2 + 2$.

****

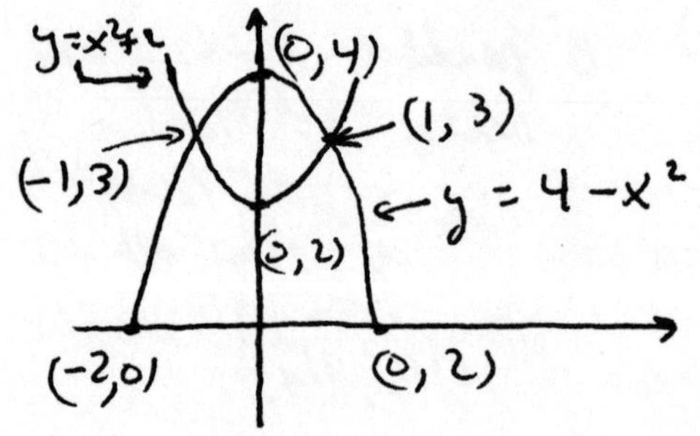

Pts. of INTERSECTION

$$4 - x^2 = x^2 + 2$$
$$2 = 2x^2$$
$$1 = x^2$$
so $x = \pm 1$

AREA BETWEEN CURVES IS:

$$A = \int_{-1}^{1} [(4 - x^2) - (x^2 + 2)]\,dx = \int_{-1}^{1} (2 - 2x^2)\,dx$$

$$= 2x - \frac{2}{3}x^3 \Big|_{-1}^{1} = (2 - 2/3) - (-2 + 2/3)$$

$$= 4/3 + 4/3 = 8/3 \text{ square units}$$

NOTE: BY SYMMETRY OF THE REGION

$$A = 2\int_0^1 (2 - 2x^2)\,dx = 8/3 \text{ square units}$$

6-5

Find the area bounded by the graphs of $y = \sqrt{x}$,

$y = \frac{5-x}{4}$, $y = \frac{3x-8}{2}$

****

$$\sqrt{x} = \frac{5-x}{4}$$
$$4\sqrt{x} = 5-x$$
$$x + 4\sqrt{x} - 5 = 0$$
$$(\sqrt{x}+5)(\sqrt{x}-1) = 0$$
$$x = 1$$
$$y = 1$$

$$\sqrt{x} = \frac{3x-8}{2}$$
$$2\sqrt{x} = 3x-8$$
$$3x - 2\sqrt{x} - 8 = 0$$
$$(3\sqrt{x}+4)(\sqrt{x}-2) = 0$$
$$x = 4$$
$$y = 2$$

$$\frac{5-x}{4} = \frac{3x-8}{2}$$
$$5-x = 6x-16$$
$$x = 3$$
$$y = \frac{1}{2}$$

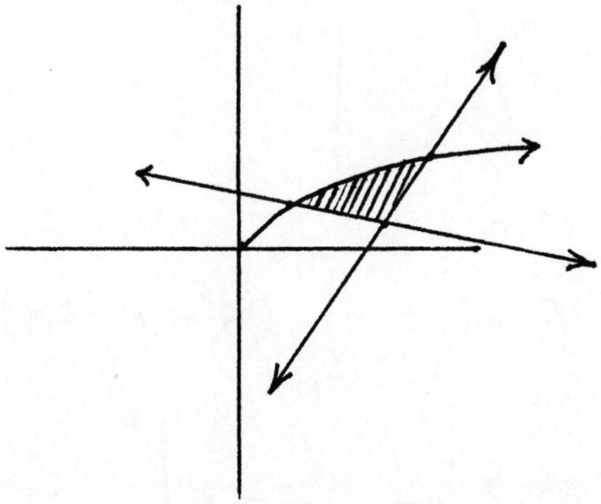

$$\text{AREA} = \int_1^3 \left(\sqrt{x} - \frac{5-x}{4}\right)dx + \int_3^4 \left(\sqrt{x} - \frac{3x-8}{2}\right)dx$$

$$= \left[\frac{2}{3}x^{3/2} - \frac{5}{4}x + \frac{x^2}{8}\right]_1^3 + \left[\frac{2}{3}x^{3/2} - \frac{3}{4}x^2 + 4x\right]_3^4$$

$$= 2\sqrt{3} - \frac{15}{4} + \frac{9}{8} - \frac{2}{3} + \frac{5}{4} - \frac{1}{8} + \frac{16}{3} - 12 + 16 - 2\sqrt{3} + \frac{27}{4} - 12$$

$$= \frac{23}{12}$$

6-6

Find the area of the finite region lying between $y = x^3 - x^2 - 2x$ and $y = 0$.

****

We first obtain a rough sketch of this curve to determine where the curve lies above and below the x-axis.

$$y = x^3 - x^2 - 2x = x(x^2 - x - 2) = x(x-2)(x+1)$$

$y = 0$ if $x = 0, -1$ or 2

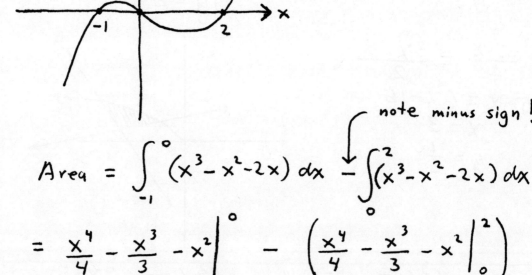

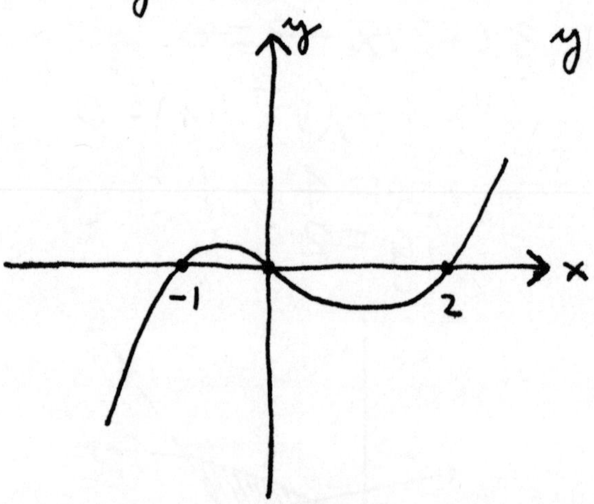

note minus sign!

$$\text{Area} = \int_{-1}^{0} (x^3 - x^2 - 2x)\,dx - \int_{0}^{2} (x^3 - x^2 - 2x)\,dx$$

$$= \frac{x^4}{4} - \frac{x^3}{3} - x^2 \Big|_{-1}^{0} - \left(\frac{x^4}{4} - \frac{x^3}{3} - x^2 \Big|_{0}^{2} \right)$$

$$= 0 - (1/4 + 1/3 - 1) - (4 - 8/3 - 4 - 0)$$

$$= 5/12 + 8/3 = 37/12 \text{ sq. units}$$

6-7

Find the area of the region bounded by $f(x) = x^2 + 2x-1$ and $g(x) = x^3-1$

****

To determine intersection points of the two graphs, set $f(x) = g(x)$:

$$x^2 + 2x - 1 = x^3 - 1 \quad \text{or}$$

$$x^3 - x^2 - 2x = 0 \quad \text{or}$$

$$x(x-2)(x+1) = 0 \quad \text{or} \quad x = 0, 2, -1.$$

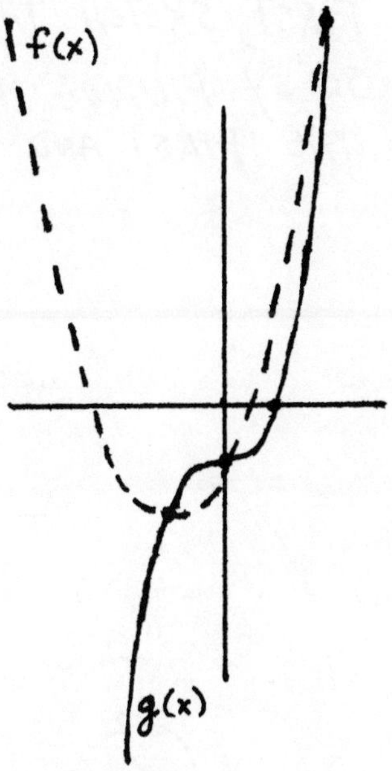

From the graph we see that g(x) is above f(x) from -1 to 0 and below from 0 to 2. Consequently, the area of the bounded region is given by

$$\int_{-1}^{0} (g(x) - f(x))\,dx + \int_{0}^{2} (f(x) - g(x))\,dx =$$

$$\int_{-1}^{0} [(x^3-1) - (x^2+2x-1)]\,dx + \int_{0}^{2} [(x^2+2x-1) - (x^3-1)]\,dx =$$

$$\int_{-1}^{0} (x^3 - x^2 - 2x)\,dx + \int_{0}^{2} (x^2 + 2x - x^3)\,dx =$$

$$\frac{x^4}{4} - \frac{x^3}{3} - x^2 \Big|_{-1}^{0} + \frac{x^3}{3} - \frac{x^4}{4} + x^2 \Big|_{0}^{2} = \frac{37}{12}.$$

6-8

Find the area of the region bounded by the x-axis and the graph of $y = 6x(x - 2)^2$.

****

FIRST, SKETCH THE GRAPH. GENERALLY, IT IS BEST TO DO SO BY FINDING INTERCEPTS AND USING INFORMATION FROM THE FIRST AND SECOND DERIVATIVES.

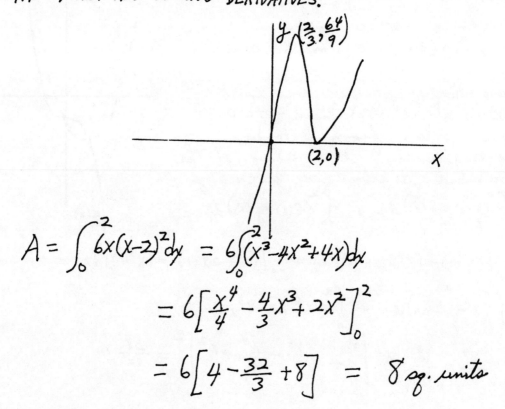

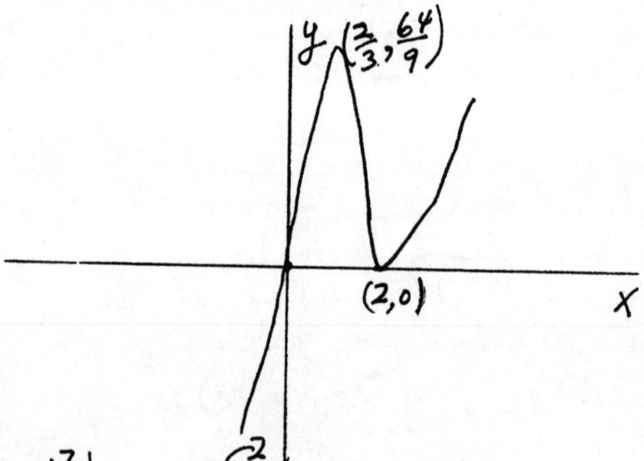

$$A = \int_0^2 6x(x-2)^2\,dx = 6\int_0^2 (x^3 - 4x^2 + 4x)\,dx$$

$$= 6\left[\frac{x^4}{4} - \frac{4}{3}x^3 + 2x^2\right]_0^2$$

$$= 6\left[4 - \frac{32}{3} + 8\right] = 8 \text{ sq. units}$$

6-9

Find the area in the plane bounded by $x = y^2 - 7$ and $x = y - 1$.

****

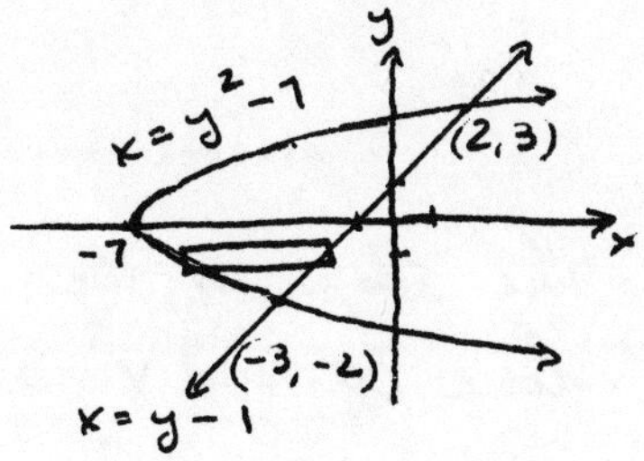

$x = y^2 - 7$ is a parabola that opens to the right and has vertex $(-7,0)$. $x = y-1$ is a line. Solve the equations simultaneously to find the points of intersection $(2,3)$ and $(-3,-2)$. We will integrate on y. A typical rectangle is shown. Its length is

$$x(\text{line}) - x(\text{parabola}) = y - 1 - (y^2 - 7).$$

So

$$\text{Area} = \int_{-2}^{3} [y - 1 - (y^2 - 7)]\, dy = \int_{-2}^{3} (y + 6 - y^2)\, dy$$

$$= \left(\frac{y^2}{2} + 6y - \frac{y^3}{3}\right) \Big|_{-2}^{3}$$

$$= \frac{3^2}{2} + 6\cdot 3 - \frac{3^3}{3} - \left[\frac{(-2)^2}{2} + 6(-2) + \frac{(-2)^3}{3}\right]$$

$$= \frac{9}{2} + 18 - 9 - \left(2 - 12 + \frac{8}{3}\right) = \frac{125}{6} = 20.83 \text{ square units}$$

Using x as the variable of integration requires two integrals, the first with limits -7 and -3, the second with limits -3 and 2. The integrals are

$$\int_{-7}^{-3} \left[\sqrt{x+7} - (-\sqrt{x+7})\right] dx + \int_{-3}^{2} \left[\sqrt{x+7} - (x+1)\right] dx$$

and the sum evaluates to $\frac{125}{6}$ as before.

6-10

Let R be the region bounded by: $y = x^3$, the tangent to $y = x^3$ at (1,1), and the x-axis. Find the area of R integrating: a) with respect to x, b) with respect to y.

****

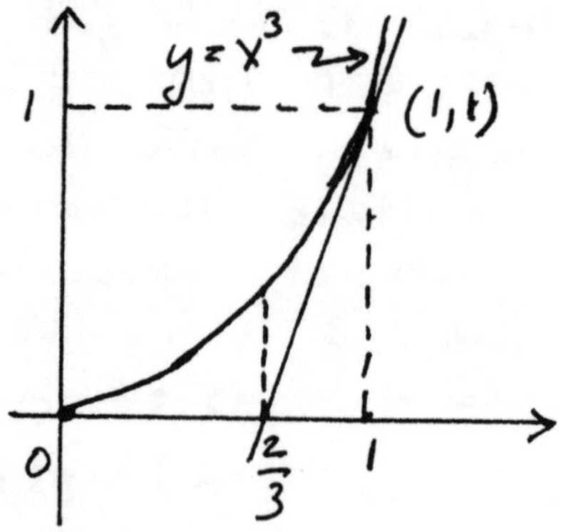

the slope of the tangent line to $y=x^3$ at (1,1) is:

$$\left.\frac{dy}{dx}\right]_{x=1} = \left.3x^2\right]_{x=1} = 3$$

Hence the equation of the tangent line is:

$$y-1 = 3(x-1) \quad \text{or} \quad y = 3x-2$$

the x-intercept of the tangent is given by:

$$3x-2=0 \quad \text{or} \quad x = 2/3 \quad \text{(see sketch above)}$$

a) to find the area A of R integrating with respect to x, we need to break the integration into 2 parts: $0 \le x < \frac{2}{3}$ and $\frac{2}{3} \le x \le 1$

$$A = \int_0^{2/3} x^3\,dx + \int_{2/3}^1 (x^3 - (3x-2))\,dx$$

$$= \int_0^{2/3} x^3\,dx + \int_{2/3}^1 (x^3 - 3x + 2)\,dx$$

$$= \left.\frac{1}{4}x^4\right]_0^{2/3} + \left(\frac{1}{4}x^4 - \frac{3}{2}x^2 + 2x\right]_{2/3}^1$$

$$= \frac{1}{4}\left(\frac{2}{3}\right)^4 + \left(\frac{1}{4} - \frac{3}{2} + 2\right) - \left(\frac{1}{4}\left(\frac{2}{3}\right)^4 - \frac{3}{2}\cdot\left(\frac{2}{3}\right)^2 + 2\cdot\frac{2}{3}\right)$$

$= \frac{1}{12}$

b) To find the area A of R integrating with respect to y, we solve

$y = x^3$ and $y = 3x - 2$ for x to get:

$x = y^{1/3}$ and $x = \frac{1}{3}y + \frac{2}{3}$

From the sketch we see that

$$A = \int_0^1 \left(\left(\frac{1}{3}y + \frac{2}{3}\right) - y^{1/3}\right) dy$$

$$= \left(\frac{1}{6}y^2 + \frac{2}{3}y - \frac{3}{4}y^{4/3}\right]_0^1$$

$$= \frac{1}{6} + \frac{2}{3} - \frac{3}{4} = \frac{1}{12}$$

Note that the answers in (a) and (b) agree!

6-11

Find the area between the graphs of $y = x^2 - 2x$ and $y = -2x^2 + 3x + 2$.

****

Points of intersection: $x^2 - 2x = -2x^2 + 3x + 2$; $3x^2 - 5x - 2 = 0$; $(3x + 1)(x - 2) = 0$; $x = \frac{-1}{3}, 2$.

$$\text{Area} = \left| \int_{-1/3}^{2} \left((-2x^2 + 3x + 2) - (x^2 - 2x)\right) dx \right| = \left| \int_{-1/3}^{2} (-3x^2 + 5x + 2)\, dx \right|$$

$$= \left| -x^3 + \frac{5}{2}x^2 + 2x \Big]_{-1/3}^{2} \right| = \left| (-8 + 10 + 4) - \left(\frac{1}{27} + \frac{5}{18} - \frac{2}{3}\right) \right|$$

$$= \left| 6 - \frac{19}{54} \right| = 5\frac{35}{54}.$$

6-12

Find the area between the curves $y = x^2$ and $y = x + 1$.

****

To find the limits of integration we solve simultaneously, $x^2 = x + 1$, or $x^2 - x - 1 = 0$, or $x = \frac{1 \pm \sqrt{5}}{2}$. Since the line is above the parabola, the area is given by

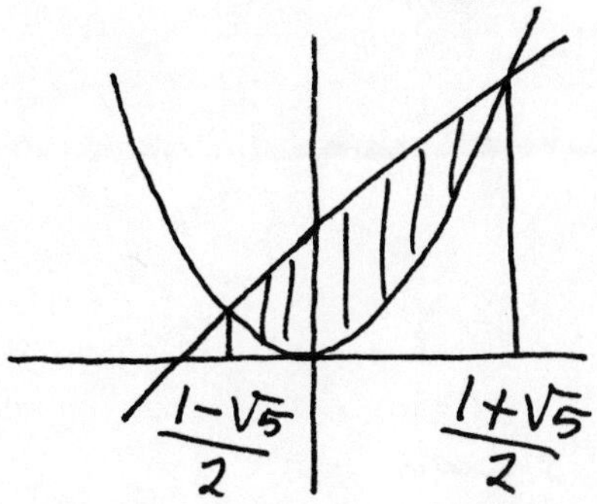

$$\int_{\frac{1-\sqrt{5}}{2}}^{\frac{1+\sqrt{5}}{2}} (x + 1 - x^2)\,dx = \frac{x^2}{2} + x - \frac{x^3}{3} \Bigg|_{\frac{1-\sqrt{5}}{2}}^{\frac{1+\sqrt{5}}{2}}$$

$$= \left(\frac{1+\sqrt{5}}{2}\right)^2 \Big/ 2 + \frac{1+\sqrt{5}}{2} - \left(\frac{1+\sqrt{5}}{2}\right)^3 \Big/ 3$$

$$- \left(\frac{1-\sqrt{5}}{2}\right)^2 \Big/ 2 - \frac{1-\sqrt{5}}{2} + \left(\frac{1-\sqrt{5}}{2}\right)^3 \Big/ 3$$

$$= \frac{1}{8}\left[1 + 2\sqrt{5} + 5 - 1 + 2\sqrt{5} - 5\right] + \sqrt{5} + \frac{1}{24}\left[-1 - 3\sqrt{5} - 15 - 5\sqrt{5} + 1 - 3\sqrt{5} + 15 - 5\sqrt{5}\right]$$

$$= \frac{\sqrt{5}}{2} + \sqrt{5} - \frac{2\sqrt{5}}{3}$$

$$= \frac{5\sqrt{5}}{6}.$$

VOLUME OF A SOLID OF REVOLUTION
CIRCULAR-DISK METHOD

6-13 ■■

The volume of a sphere of radius r is $4\pi r^3/3$. Verify this formula by revolving the circle $x^2 + y^2 = r^2$ about the x axis.

****

The equation of the circle can be written as
$$y^2 = r^2 - x^2.$$

A typical disk is shown.

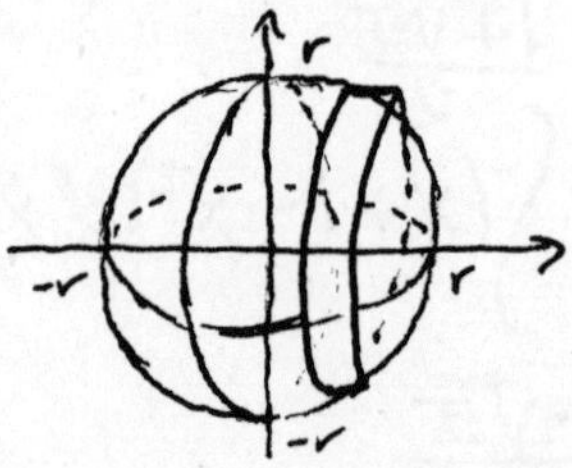

Its radius is y. By symmetry, we can rotate the portion of the curve from 0 and r and double the result to find the volume. Remember that r is a constant.

$$\text{Volume} = 2 \int_0^r \pi y^2 dx = 2\pi \int_0^r (r^2 - x^2) dx$$

$$= 2\pi \left(r^3 x - \frac{1}{3} x^3 \right) \Big|_0^r = 2\pi \left(r^3 - \frac{1}{3} r^3 \right)$$

$$= 2\pi \left(\frac{2}{3} r^3 \right) = \frac{4}{3} \pi r^3$$

6-14

Find the volume of the solid generated by revolving about the line y = 4 the smaller region bounded by the curve $x^2 = 4y$ and the lines x = 2 and y = 4.

****

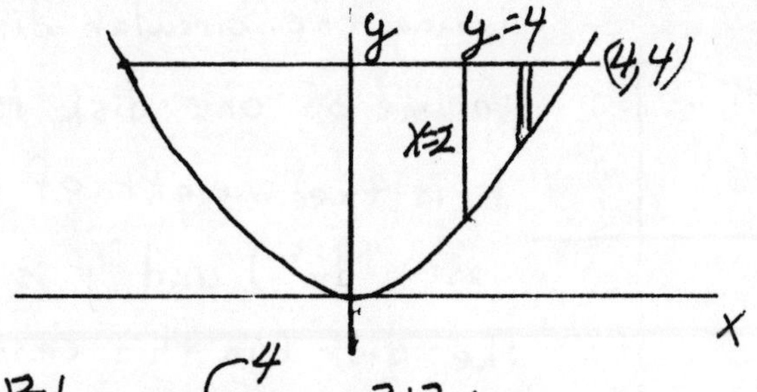

$$V = \pi\int_{x=2}^{x=4}(4-y)^2\,dx = \pi\int_2^4\left(4-\frac{x^2}{4}\right)^2 dx$$

$$= \pi\int_2^4\left(16-2x^2+\frac{x^4}{16}\right)dx$$

$$= \pi\left[16x-\frac{2}{3}x^3+\frac{x^5}{80}\right]_2^4$$

$$= \pi\left[64-\frac{128}{3}+\frac{64}{5}-32+\frac{16}{3}-\frac{2}{5}\right] = \frac{106}{15}\pi$$

CUBIC UNITS

NOTE!
THE CYLINDRICAL-SHELL METHOD MAY ALSO BE USED.

6-15

Consider the region in the xy plane between x=0 and x=$\pi/2$, bounded by y=0 and y= sin x . Find the volume of the solid generated by revolving this region about the x-axis.

****

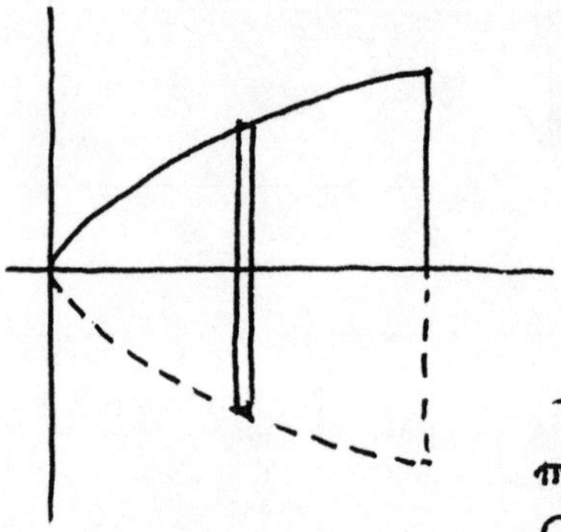

Using the circular disk method, the volume of one disk is $\pi r^2 h$, where h is the height of the disk (in this case "dx") and r is the radius of the disk (in this case "sin x").

$$\therefore \text{Volume} = \int_0^{\pi/2} \pi (\sin x)^2\, dx = \pi \int_0^{\pi/2} \sin^2 x\, dx$$

$$= \pi \int_0^{\pi/2} \frac{1-\cos 2x}{2}\, dx = \pi \left[\frac{x}{2} - \frac{\sin 2x}{4} \right]_0^{\pi/2}$$

$$= \pi \left[\left(\frac{\pi/2}{2} - \frac{\sin \pi}{4} \right) - \left(\frac{0}{2} - \frac{\sin 0}{2} \right) \right] = \pi \left[\left(\frac{\pi}{4} - 0 \right) - 0 \right]$$

$$= \pi^2/4 \text{ cubic units}$$

6-16

Find the volume of the solid formed when the region bounded by the curves $y = x^3 + 1$, $x = 1$ and $y = 0$ is rotated about the x-axis.

****

Since the axis of rotation is a boundary line for the region, the circular disk method is the best method for finding the Volume. So we should draw the representative rectangle perpendicular to the axis of rotation so that as the region is rotated a cylinder is formed. See the diagram below.

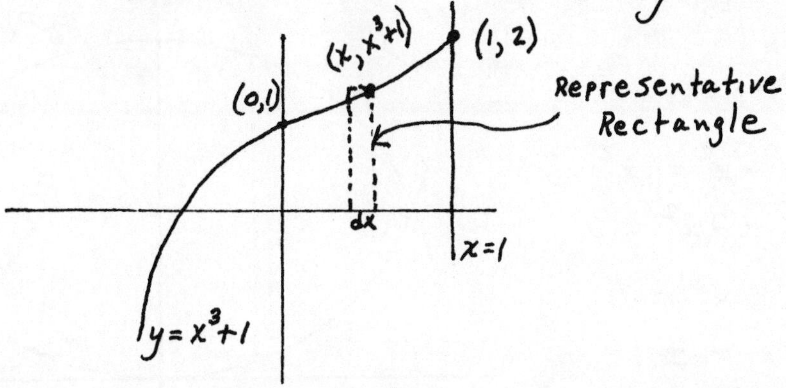

As the region is rotated, the representative rectangle of width dx and height (x^3+1) forms a circular cylinder whose Volume is $\pi r^2 h$ or $\pi(x^3+1)^2 dx$.

Hence the volume of the region is given by

$$V = \int_0^1 \pi (x^3+1)^2 dx$$

$$= \int_0^1 \pi(x^6 + 2x^3 + 1) dx$$

$$= \pi \left[\frac{x^7}{7} + \frac{1}{2}x^4 + x \right]_0^1$$

$$= \frac{23}{14}\pi \text{ cubic units}$$

VOLUME OF A SOLID OF REVOLUTION CIRCULAR-RING METHOD

6-17

Find the volume of the solid obtained when the region enclosed by $y = x^2$ and $y = 2 - x^2$ is revolved about the x-axis.

****

Find intersection:

$$2 - x^2 = x^2$$
$$2 - 2x^2 = 0$$
$$1 - x^2 = 0$$
$$x = \pm 1$$

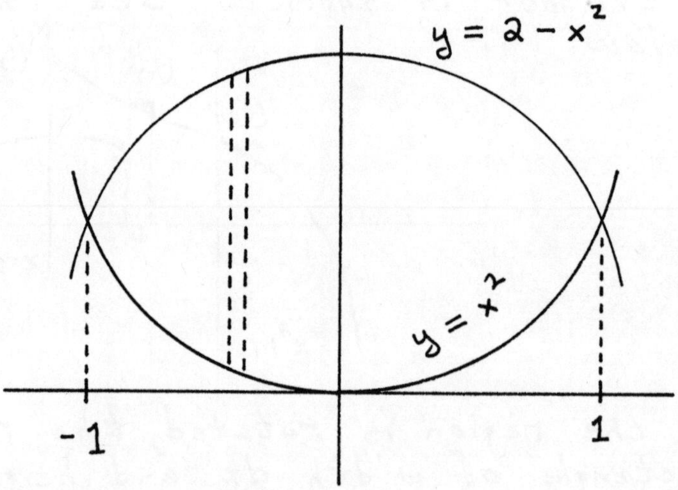

Use the method of "washers":

$$\text{Vol.} = \pi \int_{-1}^{1} [(2-x^2)^2 - (x^2)^2]\,dx$$

$$= \pi \int_{-1}^{1} (4 - 4x^2 + x^4 - x^4)\,dx$$

$$= \pi \int_{-1}^{1} (4 - 4x^2)\,dx = \pi\left(4x - \frac{4x^3}{3}\right)\Bigg|_{-1}^{1}$$

$$= \pi[(8/3) - (-8/3)] = \frac{16\pi}{3} \text{ cubic units}$$

6-18

Find the volume of the solid generated by revolving about the line $y = -1$ the region bounded by the graphs of the following equations

$$y = x^2 - 4x + 5 \text{ and } y = 5 - x$$

****

The figure region, the rectangular element of area and the points of intersection, (0,5) and (3,2).

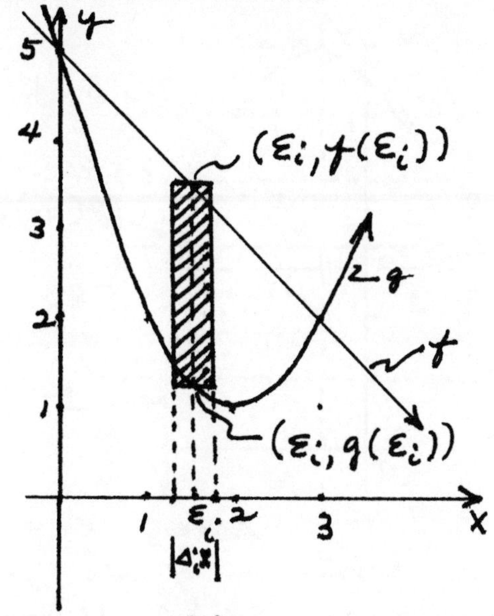

The measure of the volume of the circular-ring is

$$\Delta_i V = \pi\left\{[f(\varepsilon_i)+1]^2 - [g(\varepsilon_i)+1]^2\right\}\Delta_i x$$

$$V = \lim_{\|\Delta\|\to 0}\sum_{i=1}^{n}\Delta_i V = \pi\int_0^3\left\{[f(x)+1]^2 - [g(x)+1]^2\right\}dx$$

$$= \pi\int_0^3[(6-x)^2 - (x^2-4x+6)^2]\,dx$$

$$= \pi\int_0^3(-x^4+8x^3-27x^2+36x)\,dx$$

$$= \pi\left(-\tfrac{1}{5}x^5+2x^4-9x^3+18x^2\right)\Big|_0^3$$

$$= \frac{162\pi}{5} \text{ cu. units}$$

6-19

A curve described by the equation $(x - 1)^2 = 32 - 3y$ is rotated about the line $x = 1$ to generate a solid of revolution. Find the volume of this solid for the region bounded by $x = 1$, $y = 1$, and $y = 4$ and to the right of $x = 1$.

****

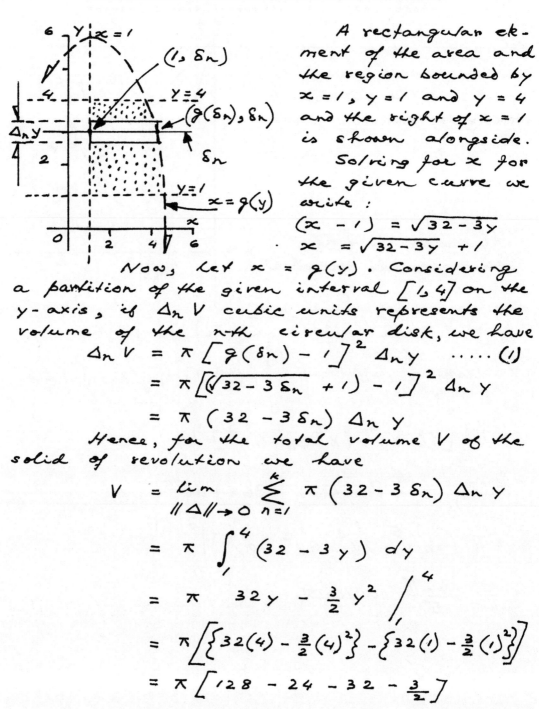

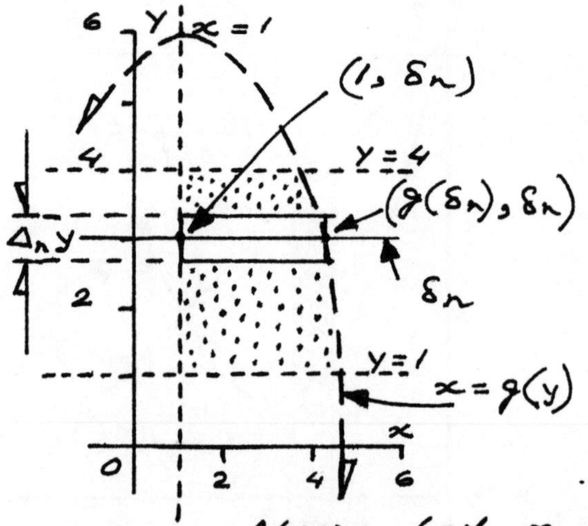

A rectangular element of the area and the region bounded by $x = 1$, $y = 1$ and $y = 4$ and the right of $x = 1$ is shown alongside.

Solving for x for the given curve we write:

$$(x - 1) = \sqrt{32 - 3y}$$

$$x = \sqrt{32 - 3y} + 1$$

Now, let $x = g(y)$. Considering a partition of the given interval $[1, 4]$ on the y-axis, if $\Delta_n V$ cubic units represents the volume of the n-th circular disk, we have

$$\Delta_n V = \pi [g(\delta_n) - 1]^2 \Delta_n y \quad \ldots\ldots (1)$$

$$= \pi [(\sqrt{32 - 3\delta_n} + 1) - 1]^2 \Delta_n y$$

$$= \pi (32 - 3\delta_n) \Delta_n y$$

Hence, for the total volume V of the solid of revolution we have

$$V = \lim_{\|\Delta\| \to 0} \sum_{n=1}^{k} \pi (32 - 3\delta_n) \Delta_n y$$

$$= \pi \int_1^4 (32 - 3y)\, dy$$

$$= \pi \; 32y - \frac{3}{2} y^2 \Big|_1^4$$

$$= \pi \left[\left\{ 32(4) - \frac{3}{2}(4)^2 \right\} - \left\{ 32(1) - \frac{3}{2}(1)^2 \right\} \right]$$

$$= \pi \left[128 - 24 - 32 - \frac{3}{2} \right]$$

$$= \frac{141}{2}\pi$$

Hence, the volume of the solid of revolution is $\frac{141}{2}\pi$ cubic units.

6-20

Find the volume of the indicated area revolved about the x-axis.

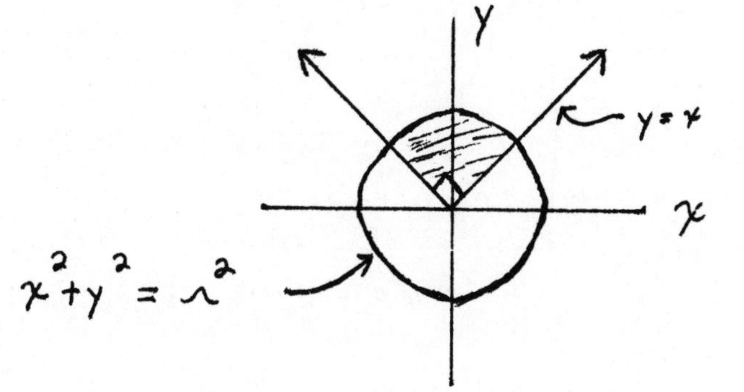

****

We will use the washer method i.e. the difference between two disks.

$$\text{Volume} = 2\pi\int_0^{\frac{r}{\sqrt{2}}} ((r^2-x^2)-x^2)\,dx = 2\pi\int_0^{\frac{r}{\sqrt{2}}} r^2-2x^2\,dx =$$

$$2\pi\left\{ r^2x - \frac{2x^3}{3}\right|_0^{\frac{r}{\sqrt{2}}} = 2\pi\left\{\left(\frac{r^3}{\sqrt{2}} - \frac{2r^3}{6\sqrt{2}}\right)-(0-0)\right\} =$$

$$\sqrt{2}\pi\left\{\frac{2r^3}{3}\right\} = \frac{2\sqrt{2}r^3\pi}{3}.$$

6-21

Find the volume of the solid of revolution obtained by revolving the region bounded by $y=x^2$, the x-axis, $x=2$ around the line $y = -1$.

****

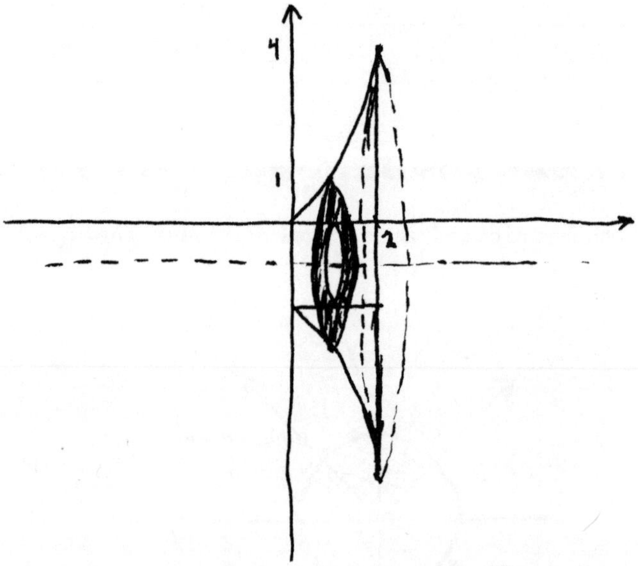

Note $R = y+1 = x^2+1$

$$\text{Area of circular ring} = \pi R^2 - \pi r^2$$
$$= \pi(x^2+1)^2 - \pi(1)^2$$
$$= \pi(x^4+2x^2+1) - \pi$$
$$= \pi(x^4+2x^2+1-1) = \pi(x^4+2x^2)$$

$$\Rightarrow \text{Volume} = \pi\int_0^2 (x^4+2x^2)\,dx = \pi\left(\frac{x^5}{5}+\frac{2x^3}{3}\right]_0^2$$

$$= \pi\left[\frac{32}{5}+\frac{16}{3}-0\right] = \pi\left(\frac{96+80}{15}\right) = \frac{176}{15}\pi \text{ cubic units}$$

6-22

Consider the plane region bounded by the curves given by

$y = x^2 + 1$ and $y = x + 3$

Find the volume of the solid of revolution generated by revolving the region around the x-axis.

****

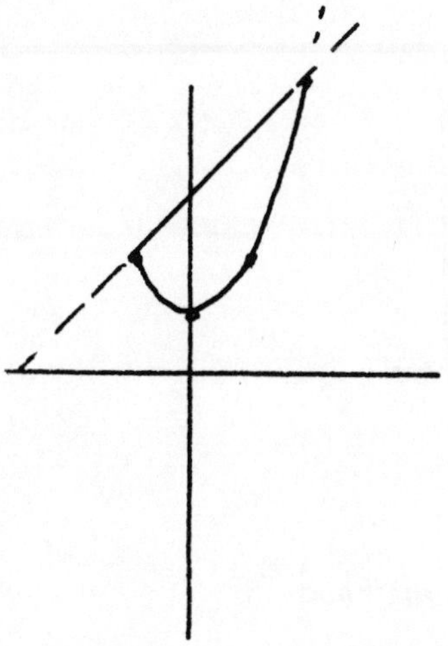

$y = x^2+1$ and $y = x+3$ intersect where

$$x^2+1 = x+3$$

$\Rightarrow \quad x^2 - x - 2 = 0$

$\Rightarrow \quad (x-2)(x+1) = 0$

or $\boxed{x=2,\ y=5}$ OR $\boxed{x=-1,\ y=2}$

When rotating around the x-axis, vertical cross-sections are circular rings.

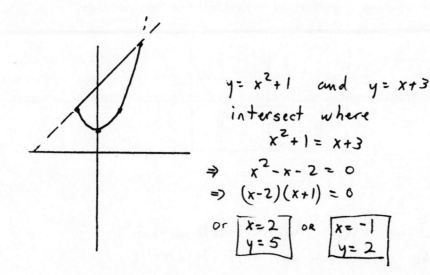

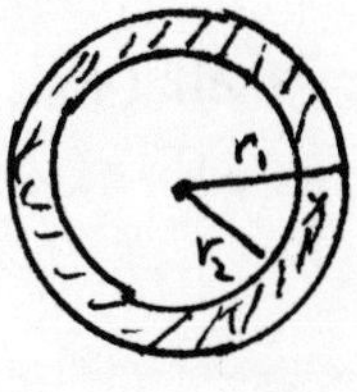

$r_1 = x+3$

$r_2 = x^2+1$

cross-sectional area $= \pi(r_1^2 - r_2^2)$

Thus the volume

$$V = \int_{-1}^{2} \pi\left(r_1^2 - r_2^2\right) dx$$

$$= \int_{-1}^{2} \pi\left((x+3)^2 - (x^2+1)^2\right) dx$$

$$= \pi \int_{-1}^{2} (-x^4 - x^2 + 6x + 8)\, dx = \pi\left(-\tfrac{1}{5}x^5 - \tfrac{1}{3}x^3 + 3x^2 + 8x\right)\Big|_{-1}^{2}$$

$$= \pi\left[\left(-\tfrac{32}{5} - \tfrac{8}{3} + 12 + 16\right) - \left(\tfrac{1}{5} + \tfrac{1}{3} + 3 - 8\right)\right]$$

$= \boxed{\frac{117}{5}\pi}$ cubic units

VOLUME OF A SOLID OF REVOLUTION CYLINDRICAL-SHELL METHOD

6-23

The region in the xy plane bounded by $y=x^2$, $y=0$ and $x=1$ is revolved around the y-axis. Find the volume of the solid generated.

****

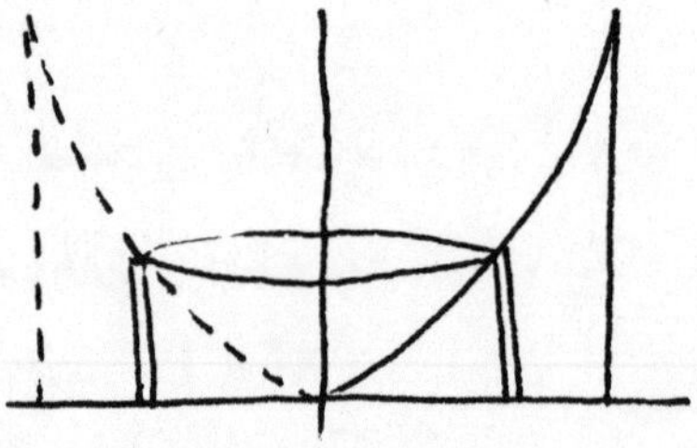

Using the cylindrical shell method, the volume of one shell is $2\pi rht$, where t is the thickness of the shell (this time "dx"), h is the height of the shell (this time "x^2") and r is the radius of the shell (this time "x").

$$\therefore \text{Volume} = \int_0^1 2\pi(x)(x^2)\,dx = 2\pi\int_0^1 x^3\,dx = 2\pi \frac{x^4}{4}\Big|_0^1$$

$$= 2\pi\left[\frac{1}{4} - 0\right] = \pi/2 \text{ cubic units}$$

6-24

Given A: $y^2 = 2(x - 1)$ and B: $y^2 = 4(x - 2)$.
Sketch the curves labelling all points of intersection.
Find the volume of the solid of revolution generated by revolving the bounded region about the x-axis.
Find the volume of the solid of revolution generated by revolving the bounded region about the y-axis.

****

A: parabola, opens right, vertex (1, 0)

B: parabola opens right, vertex (2, 0)

pts. of intersection $2(x-1) = 4(x-2) \Rightarrow 2x-2 = 4x-8$
$3 = x$

substitute $x=3$ into A or B $\Rightarrow y = \pm 2$

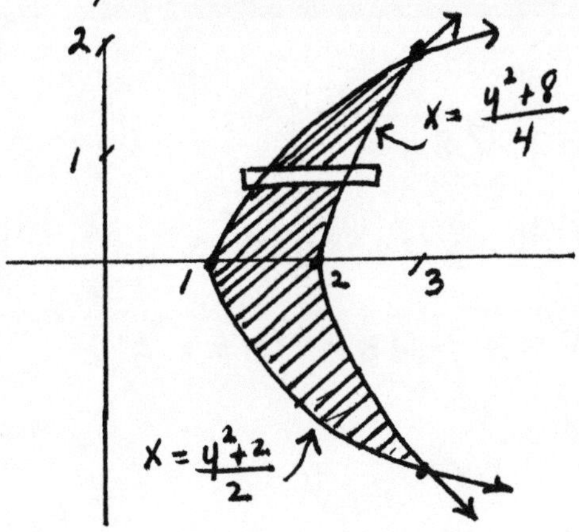

Around x-axis:
Use method of shells on region in Q I.

$$V = \int_0^2 2\pi(y)\left(\frac{y^2+8}{4} - \frac{y^2+2}{2}\right) dy$$

$$= 2\pi \int_0^2 \left(-\frac{1}{4}y^3 + y\right) dy$$

$$= 2\pi \left(-\frac{y^4}{16} + \frac{y^2}{2}\right)\Big|_0^2 = 2\pi \text{ cubic units}$$

Around y-axis:
Use method of circular rings

$$V = \int_{-2}^{2} \left[\pi\left(\frac{y^2+8}{4}\right)^2 - \pi\left(\frac{y^2+2}{2}\right)^2\right] dy = \pi \int_{-2}^{2} \left(-\frac{3}{16}y^4 + 3\right) dy$$

$$= \pi\left(-\frac{3y^5}{80} + 3y\right)\Big|_{-2}^{2} = \frac{48\pi}{5} \text{ cubic units}$$

6-25

Use cylindrical shells to find the volume of the solid obtained by revolving around the y-axis the region bounded by the curves $y^2 = 8x$ and $x = 2$.

****

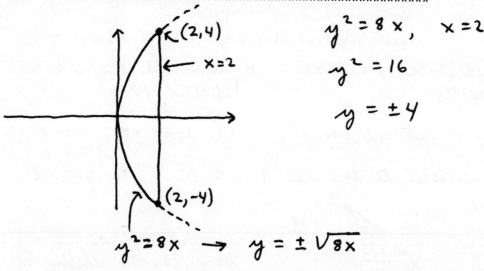

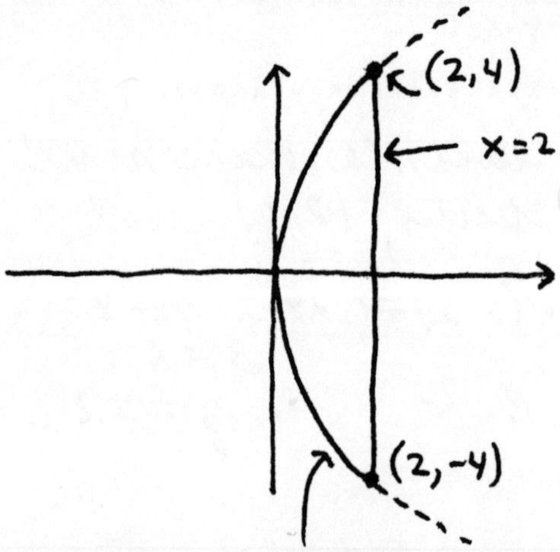

$y^2 = 8x, \quad x = 2$

$y^2 = 16$

$y = \pm 4$

$y^2 = 8x \rightarrow y = \pm\sqrt{8x}$

Using cylindrical shells, each cylinder has radius x and height $2\sqrt{8x}$, for $0 \le x \le 2$

$$\text{Volume} = \int_0^2 2\pi \underset{r}{x} (\underset{h}{2\sqrt{8x}})\, dx$$

$$= 8\sqrt{2}\,\pi \int_0^2 x^{3/2}\, dx = 8\sqrt{2}\,\pi \left. \frac{x^{5/2}}{5/2} \right|_0^2$$

$$= \frac{16\sqrt{2}\pi}{5} 2^{5/2} = \frac{128\pi}{5} \text{ cubic units}$$

6-26

A cylindrical hole has been drilled straight through the center of a sphere of radius R. Use the method of cylindrical shells to find the volume of the remaining solid if it is 6 cm high.

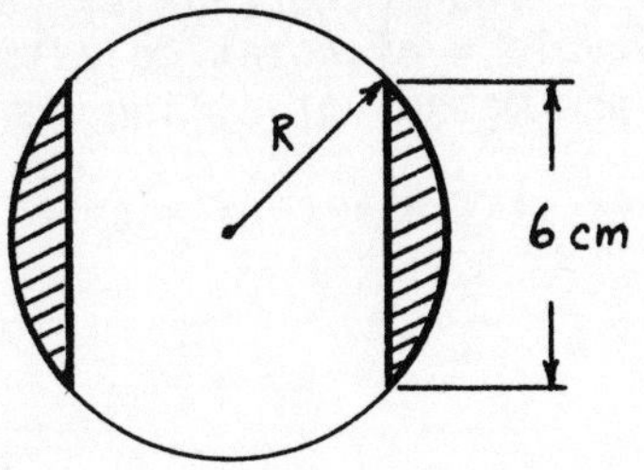

****

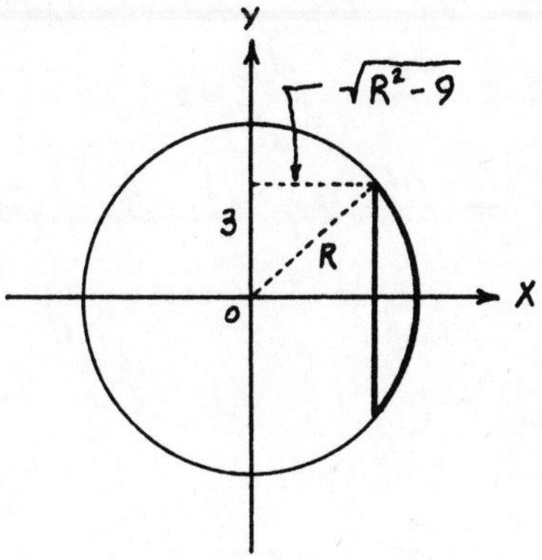

Introduce a coordinate system so that the origin is at the center of the sphere and the hole is drilled along the y-axis.

Note that the equation of the cross section of the sphere is $x^2 + y^2 = R^2 \Rightarrow y = \pm\sqrt{R^2 - x^2}$ and that the hole intersects the x-axis at $(\sqrt{R^2-9}, 0)$.

Taking advantage of symmetry, we have

$$V = 2\int_{\sqrt{R^2-9}}^{R} 2\pi x\sqrt{R^2-x^2}\,dx$$

Let $u = R^2 - x^2$, so $\frac{du}{dx} = -2x \Rightarrow dx = \frac{du}{-2x}$

When $x = R$, $u = 0$ and when $x = \sqrt{R^2-9}$, $u = 9$

So $$V = 4\pi\int_9^0 x\sqrt{u}\,\frac{du}{-2x} = -\frac{4\pi}{3}u^{3/2}\Big]_9^0 = \underline{\underline{36\pi}}.$$

Note that the volume is a constant which does not depend on R.

6-27

Let R be the region bounded by : $y = \frac{1}{x}$, $y = x^2$, $x = 0$, $y = 2$.

Suppose R is revolved around the x-axis. Set up (but do NOT evaluate) the integrals for the volume of rotation using (a) the method of circular disks /rings and (b) the method of cylindrical shells.

****

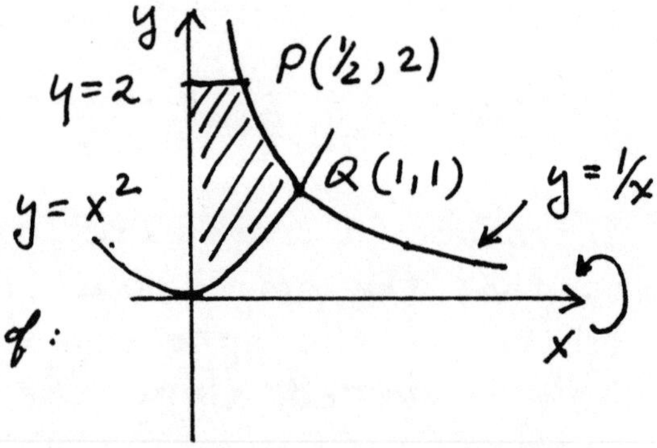

R is the hatched region in the figure.

$P = (\frac{1}{2}, 2)$ is the intersection of: $y = 2$ and $y = 1/x$

$Q = (1, 1)$ is the intersection of: $y = x^2$ and $y = 1/x$

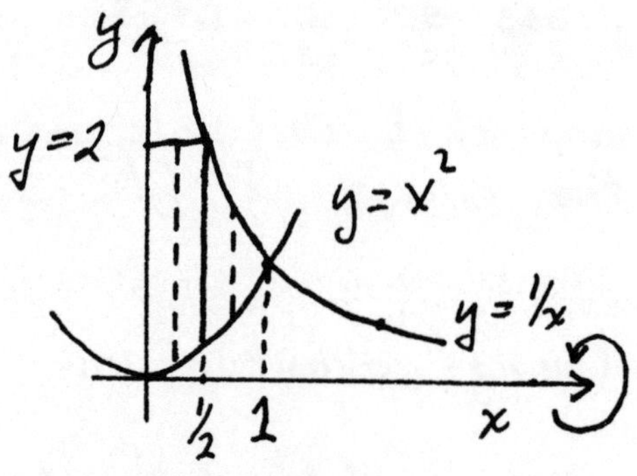

(a) For the method of circular disks/rings, we slice R perpendicular to the axis of rotation (x-axis)

The volume of a typical circular ring of thickness dx is given by:

For $0 \le x \le 1/2$ $\quad Vol_{ring} = \pi(2^2 - (x^2)^2) = \pi(4 - x^4)$

For $\frac{1}{2} \le x \le 1$ $\quad Vol_{ring} = \pi\left(\left(\frac{1}{x}\right)^2 - (x^2)^2\right) = \pi\left(\frac{1}{x^2} - x^4\right)$

Hence, the volume of rotation is given by:

$$\pi \int_0^{1/2} (4 - x^4)\, dx + \pi \int_{1/2}^{1} \left(\frac{1}{x^2} - x^4\right) dx$$

(b) For the method of cylindrical shells, we slice R parallel to the axis of rotation (x-axis)

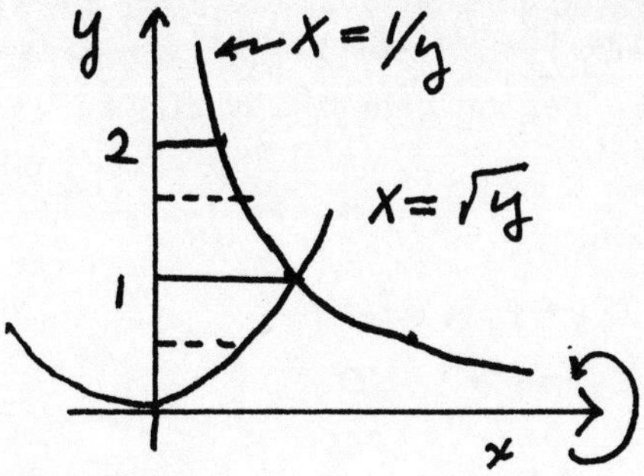

Solving $y = \frac{1}{x}$ and $y = x^2$ for x in terms of y:

$$x = \frac{1}{y}, \quad x = \sqrt{y}$$

The volume of a typical cylindrical shell of thickness dy is given by:

For $0 \leq y \leq 1$: $\quad Vol_{shell} = 2\pi y \sqrt{y}\, dy = 2\pi y^{3/2}\, dy$

For $1 \leq y \leq 2$ $\quad Vol_{shell} = 2\pi y \cdot \frac{1}{y}\, dy = 2\pi\, dy$

Hence the volume of rotation is given by:

$$2\pi \int_0^1 y^{3/2}\, dy + 2\pi \int_1^2 dy$$

6-28

Find the volume of the solid generated by revolving about the line $x = -1$, the region bounded by the curves $y = -x^2+4x-3$ and $y = 0$.

****

$$-x^2+4x-3=0$$
$$x^2-4x+3=0$$
$$(x-1)(x-3)=0$$
$$x=1,3$$

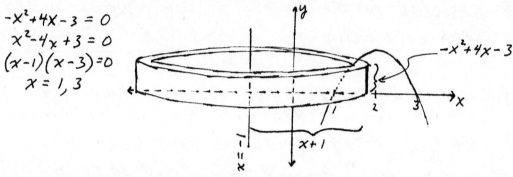

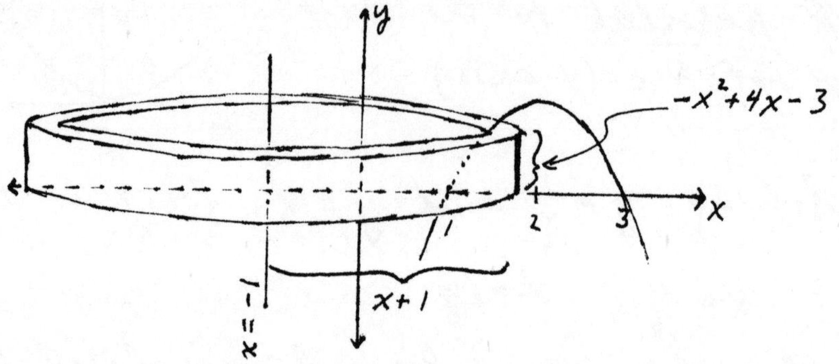

$$\int_1^3 2\pi(x+1)(-x^2+4x-3)\,dx = \int_1^3 2\pi(-x^3+3x^2+x-3)\,dx$$

$$= 2\pi\left(-\frac{1}{4}x^4+x^3+\frac{1}{2}x^2-3x\right)\Big|_1^3$$

$$= 2\pi\left[\left(-\frac{1}{4}\cdot 81+27+\frac{1}{2}\cdot 9-9\right)-\left(-\frac{1}{4}+1+\frac{1}{2}-3\right)\right]$$

$$= 2\pi\left(18-\frac{81}{4}+\frac{18}{4}+2+\frac{1}{4}-\frac{2}{4}\right)$$

$$= 2\pi\left(20-\frac{64}{4}\right)$$

$$= 8\pi \text{ cubic units}$$

6-29

Find the volume of the solid formed when the region R bounded by $y= x^3 + 1$, $x= 1$ and $y= 1$ is rotated about the line $x= -1$.

****

We choose the cylindrical shell method because the axis of rotation neither bounds nor intersects the region R. In this case we must draw the representative rectangle parallel to the axis of revolution. The formula for cylindrical shells that is needed here is given by

$$V = \int_0^1 2\pi x y \, dx$$ where x is the

radius to the inside of the cylindrical shell, $y = (x^3+1-1)$ is the height of the cylindrical shell and dx is the thickness of the shell. See the diagram below.

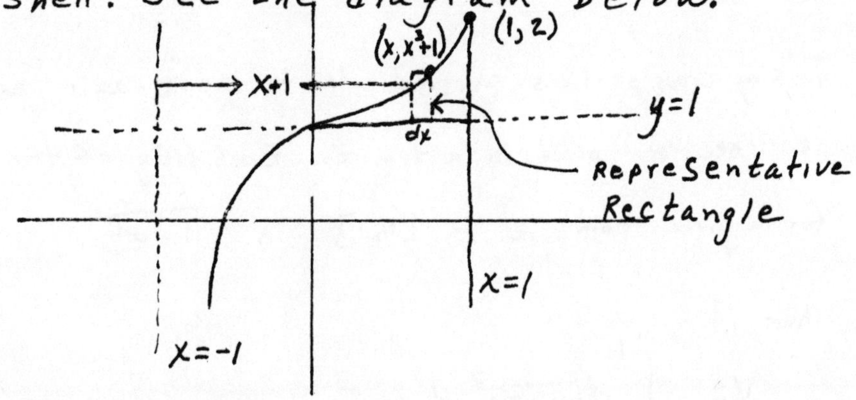

So the volume is given by

$$V = \int_0^1 2\pi(x-(-1))(x^3+1-1)\,dx$$

$$= 2\pi \int_0^1 (x^4 + x^3)\,dx$$

$$= 2\pi \left[\frac{x^5}{5} + \frac{x^4}{4}\right]_0^1$$

$$= \frac{9\pi}{10} \text{ cubic units}$$

VOLUME OF A SOLID OF REVOLUTION BY PARALLEL PLANE SECTIONS

6-30

Find the volume of one octant of the region common to two right circular cylinders of radius 1 whose axes intersect at right angles as shown below:

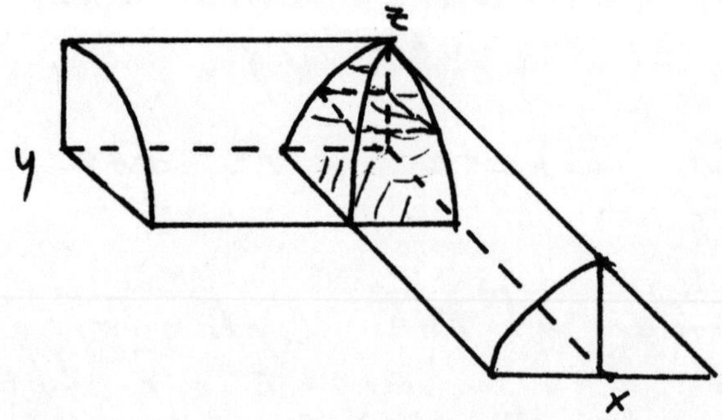

****

Taking cross sections perpendicular to the z-axis, each cross sectional piece is a square. Each side of the square for a given point z in $[0,1]$ is $\sqrt{1-z^2}$

thus:

$$V = \int_0^1 \left(\sqrt{1-z^2}\right)^2 dz$$

$$= \int_0^1 1 - z^2 \, dz$$

$$= z - \frac{z^3}{3} \Big|_0^1$$

$$= \boxed{2/3}$$

LENGTH OF A CURVE

6-31

Find the length of the curve $y = 2x^{3/2} + 1$ from $x = 0$ to $x = 4$.

****

$$L = \int_a^b \sqrt{1 + (y')^2}\, dx$$

$$y' = 2(3/2)x^{1/2} = 3x^{1/2}$$

$$(y')^2 = 9x \quad \text{so } L = \int_0^4 \sqrt{1 + 9x}\, dx$$

$$u = 1 + 9x$$
$$du = 9\,dx \Rightarrow 1/9\, du = dx$$
$$x = 4 \Rightarrow u = 37$$
$$x = 0 \Rightarrow u = 1$$

$$\text{so } L = \int_1^{37} u^{1/2} (1/9)\, du$$

$$= \frac{1}{9} (u^{3/2})(2/3) \Big]_1^{37}$$

$$= 2/27 \left(37^{3/2} - 1^{3/2} \right)$$

6-32

Find the arc length of the graph of $6xy - y^4 = 3$ from the point (19/12, 2) to the point (14/3, 3).

****

Clearly, it would be simplest to consider x as a function of y, $x = \dfrac{3+y^4}{6y} = \dfrac{1}{2y} + \dfrac{y^3}{6}$

Then $dx/dy = -\dfrac{1}{2y^2} + \dfrac{y^2}{2}$, and

$$1 + (dx/dy)^2 = 1 + \frac{1}{4y^4} - \frac{1}{2} + \frac{y^4}{4}$$

$$= \left(\frac{1}{2y^2} + \frac{y^2}{2}\right)^2$$

$$\text{Arc length} = \int_2^3 \sqrt{1+(dx/dy)^2}\, dy$$

$$= \int_2^3 \sqrt{\left(\frac{1}{2y^2} + \frac{y^2}{2}\right)^2}\, dy = \int_2^3 \left(\frac{1}{2} y^{-2} + \frac{1}{2} y^2\right) dy$$

$$= \frac{1}{2}\left[-\frac{1}{y} + \frac{y^3}{3}\right]_2^3 = \frac{1}{2}\left[-\frac{1}{3} + 9 + \frac{1}{2} - \frac{8}{3}\right]$$

$$= \frac{1}{2}\left[\frac{39}{6}\right] = \frac{13}{4} \text{ units}$$

6-33

Find the arc length of the curve $y = x^{2/3}$ from $x = 1$ to $x = 8$.

****

Recall the arc length formula $L = \int_a^b \sqrt{1 + [f'(x)]^2}\, dx$

For $f(x) = x^{2/3}$, $f'(x) = \frac{2}{3} x^{-1/3}$, so $[f'(x)]^2 = \frac{4}{9} x^{-2/3}$

and $1 + [f'(x)]^2 = 1 + \frac{4}{9} x^{-2/3} = \dfrac{9x^{2/3} + 4}{x^{2/3}}$

So, over the interval $1 \le x \le 8$

$$L = \int_1^8 \sqrt{\frac{9x^{2/3} + 4}{x^{2/3}}}\, dx = \int_1^8 \frac{\sqrt{9x^{2/3} + 4}}{x^{1/3}}\, dx$$

Let $u = 9x^{2/3} + 4$, so $\frac{du}{dx} = 6x^{-1/3} \Rightarrow dx = \frac{x^{1/3}}{6}\, du$

So $$L = \int_{x=1}^{8} \frac{\sqrt{u}}{x^{1/3}} \cdot \frac{x^{1/3}}{6}\, du$$

$$= \frac{1}{9} u^{3/2} \Big]_{x=1}^{8}$$

$$= \frac{1}{9} (9x^{2/3} + 4)^{3/2} \Big]_1^8 = \frac{1}{9}\left(40^{3/2} - 13^{3/2}\right)$$

6-34

Find the arc length of the curve defined by $y = \frac{2}{3}(x - 1)^{\frac{3}{2}}$ from $x = 1$ to $x = 4$.

****

$$y' = \frac{2}{3}\left(\frac{3}{2}\right)(x-1)^{\frac{1}{2}} = (x-1)^{\frac{1}{2}}$$

$$s = \int_1^4 \sqrt{1+(y')^2}\,dx = \int_1^4 \sqrt{1+(x-1)}\,dx$$

$$= \int_1^4 x^{\frac{1}{2}}\,dx = \frac{2}{3}x^{\frac{3}{2}}\Big|_1^4 = \frac{14}{3}$$

WORK

6-35

Find the work done in stretching a spring 6 inches beyond its natural length, if the spring constant k = 20 lb/ft.

****

$F(x) = kx$ (Hooke's Law)

$W = \int_a^b F(x)\,dx$ $\qquad$ $6 \text{ in} = \frac{1}{2}$ ft.

$$W = \int_0^{\frac{1}{2}} 20x\,dx = 10x^2\Big|_0^{\frac{1}{2}} = 10\left(\tfrac{1}{4}-0\right) = 2.5 \text{ ft-pounds}$$

6-36

A hemispherical tank with radius 8 feet is filled with water to a depth of 6 feet. Find the work required to empty the tank by pumping the water to the top of the tank.

****

$$x^2 + y^2 = 64$$
$$y = \sqrt{64 - x^2}$$

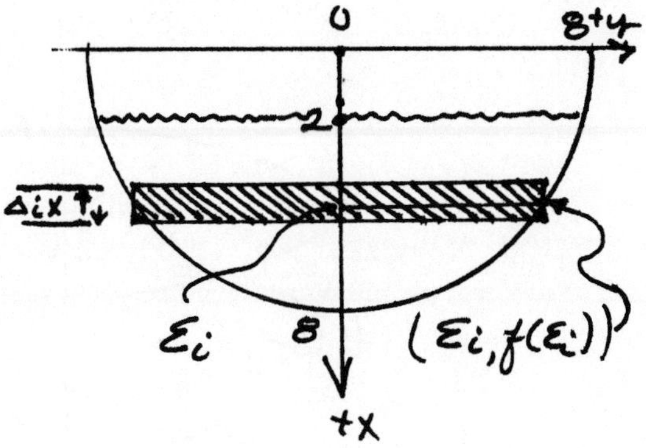

$$\Delta_i W = \pi w (64 - \varepsilon_i^2)\, \varepsilon_i \Delta_i x$$

$$W = \lim_{\|\Delta\| \to 0} \sum_{i=1}^{n} \pi w (64 - \varepsilon_i^2)\, \varepsilon_i \Delta_i x \quad \text{where}$$

w is the number of pounds in the weight of 1 ft.^{3} of water.

$$\therefore W = \pi w \int_2^8 (64 - x^2)\, x\, dx = \pi w \int_2^8 (64x - x^3)\, dx$$

$$= \pi w \left(32x^2 - \tfrac{1}{4}x^4\right)\Big|_2^8$$

$$= \pi w \left[(2048 - 1024) - (128 - 4)\right]$$

$$= 900 \pi w \text{ ft-lbs.}$$

6-37

Suppose a hemispherical tank of radius 10 feet is filled with a liquid whose density is 62 pounds per cubic foot. Find the work required to pump all of the liquid out through the top of the tank.

****

The tank may be assumed to be located as follows:

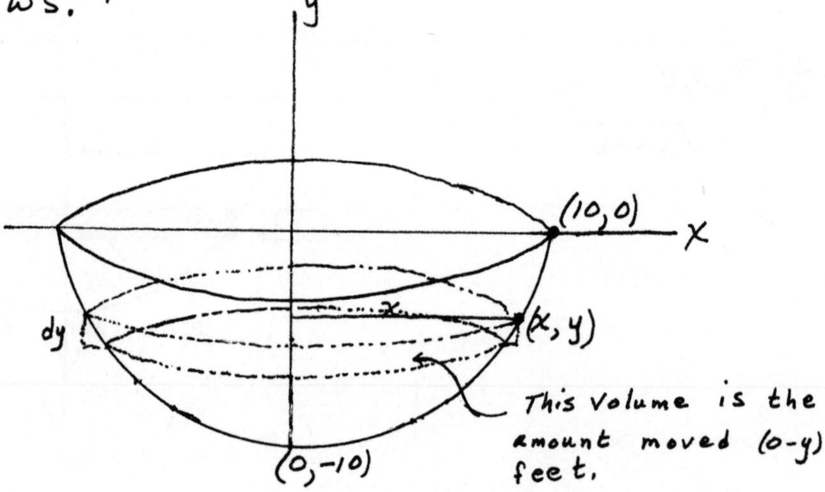

$\pi x^2 dy$ is the volume of the liquid y feet below the top of the tank. Hence the weight of liquid that is moved $(0-y)$ feet is $62\pi x^2 dy$. So the work needed to empty the tank is given by

$$\text{Work} = \int_{-10}^{0} (0-y)(62\pi x^2 dy)$$

(note: $x^2+y^2=100$)

$$= -62\pi \int_{-10}^{0} y(100-y^2)\,dy$$

$$= -62\pi \left[100\left(\frac{y^2}{2}\right) - \frac{y^4}{4} \right]_{-10}^{0}$$

$$= 155{,}000\,\pi \text{ ft-lbs.}$$

6-38

An atom is moving radially outward from the origin opposite the pull of the force $F = \frac{a}{(r-b)^3} - \frac{c}{r^6}$ where a,b,c are constants and r is the radial distance from the origin. How much work must be done to move the atom from R_1 to R_2 ?

****

$$W = \int F\,dr = \int_{R_1}^{R_2} \left(\frac{a}{(r-b)^3} - \frac{c}{r^6} \right) dr$$

$$= \int_{R_1}^{R_2} \frac{a}{(r-b)^3}\,dr - \int_{R_1}^{R_2} \frac{c}{r^6}\,dr$$

Let $u = r - b \qquad du = dr$

$$W = \int_{R_1-b}^{R_2-b} \frac{a}{u^3}\,du - \int_{R_1}^{R_2} \frac{c}{r^6}\,dr = a\int_{R_1-b}^{R_2-b} u^{-3}\,du - c\int_{R_1}^{R_2} r^{-6}\,dr$$

$$= \frac{a u^{-2}}{-2}\Bigg|_{R_1-b}^{R_2-b} - \frac{c\,r^{-5}}{-5}\Bigg|_{R_1}^{R_2} = -\frac{a}{2}\frac{1}{u^2}\Bigg|_{R_1-b}^{R_2-b} + \frac{c}{5}\frac{1}{r^5}\Bigg|_{R_1}^{R_2}$$

$$= -\frac{a}{2}\left(\frac{1}{(R_2-b)^2} - \frac{1}{(R_1-b)^2} \right) + \frac{c}{5}\left(\frac{1}{R_2^5} - \frac{1}{R_1^5} \right)$$

$$W = \text{Work} = \frac{a}{2}\left(\frac{1}{(R_1-b)^2} - \frac{1}{(R_2-b)^2} \right) + \frac{c}{5}\left(\frac{1}{R_2^5} - \frac{1}{R_1^5} \right).$$

CENTER OF MASS

6-39

A four foot rod has a density function $\rho(x) = 2x + 1$, where x is the number of feet from the left end. Where is the center of mass?

****

Set the left end of the rod at the origin. Let M be the mass and M_0 be the moments about the origin.

Then $M = \int_0^4 \rho(x)\,dx = \int_0^4 (2x+1)\,dx = x^2 + x \Big|_0^4 = 20$

and $M_0 = \int_0^4 x\rho(x)\,dx = \int_0^4 (2x^2 + x)\,dx = \frac{2}{3}x^3 + \frac{1}{2}x^2 \Big|_0^4 = \frac{152}{3}$.

Thus the center of mass is

$$\frac{M_0}{M} = \frac{\frac{152}{3}}{20} = \frac{38}{15}$$ feet from the left end.

6-40

Find the center of mass of the homogeneous lamina bounded by $y = x^2$ and $y = 4x$.

****

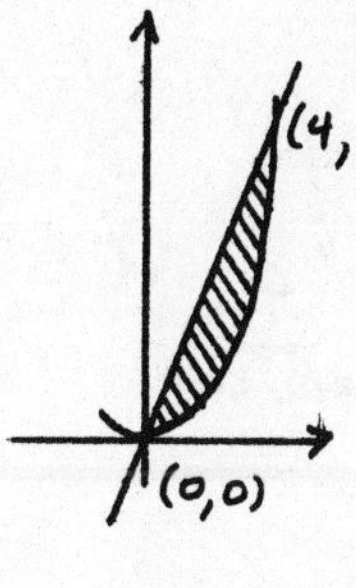

First find where the curves intersect by setting $x^2 = 4x$. Then $x^2 - 4x = 0$

$x(x-4) = 0$ so $x = 0$, $x = 4$.

The center of mass is $(\bar{x}, \bar{y})$, where $\bar{x} = \frac{M_y}{M}$ and $\bar{y} = \frac{M_x}{M}$, M = mass, M_x = moments with respect to the x axis, and M_y = moments about the y axis. "k" is the constant density.

Then $M = \int_0^4 k(4x - x^2)\,dx = k\left(2x^2 - \frac{1}{3}x^3\right)\Big|_0^4 = \frac{32}{3}k$

$$M_y = \int_0^4 kx(4x - x^2)\,dx$$

$$= k\int_0^4 (4x^2 - x^3)\,dx = k\left(\frac{4}{3}x^3 - \frac{1}{4}x^4\right)\Big|_0^4 = \frac{256}{12}k$$

$$M_x = \int_0^4 k\left(\frac{4x + x^2}{2}\right)(4x - x^2)\,dx$$ (Note: $\frac{4x+x^2}{2}$ is the average y value)

$$= \frac{k}{2}\int_0^4 (16x^2 - x^4)\,dx = \frac{k}{2}\left(\frac{16}{3}x^3 - \frac{1}{5}x^5\right)\Big|_0^4 = \frac{1024}{15}k$$

Thus $\bar{x} = \dfrac{\frac{256}{12}k}{\frac{32}{3}k} = \frac{256}{12} \cdot \frac{3}{32} = 2$ and

$$\bar{y} = \frac{\frac{1024}{15}k}{\frac{32}{3}k} = \frac{1024}{15} \cdot \frac{3}{32} = \frac{32}{5}$$

so the center of mass is at $\left(2, \frac{32}{5}\right)$.

6-41 ■■

Find the center of mass of the homogeneous lamina bounded by $y = 9 - x^2$ and the x axis.

****

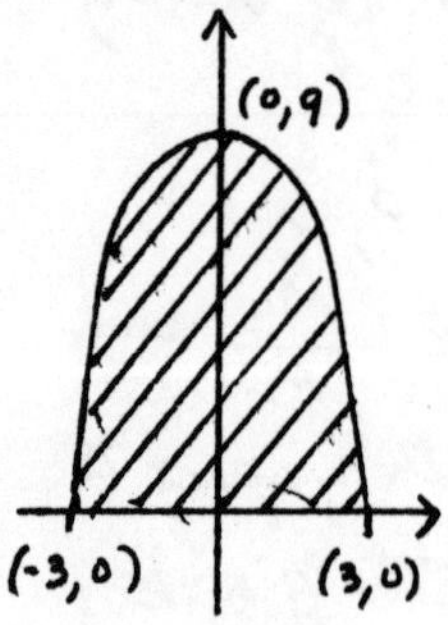

By setting $9 - x^2 = 0$ we see that the curve intersects with the x axis at -3 and 3. It is also clear that the curve is symmetrical about the y axis, so that $\bar{x}$ (the x coordinate of the center of mass) is 0.

They $\bar{y} = \dfrac{M_x}{M}$, where M is the mass and M_x is the moments about the x axis. Let k be the density.

Then $M = \displaystyle\int_{-3}^{3} k(9-x^2)dx = 2k\int_{0}^{3}(9-x^2)dx$ (using symmetry)

$$= 2k\left(9x - \tfrac{1}{3}x^3\right)\Big|_{0}^{3} = 36k$$

$$M_x = \int_{-3}^{3} k\left[\tfrac{1}{2}(9-x^2)\right](9-x^2)dx = 2\cdot\tfrac{1}{2}k\int_{0}^{3}(81-18x^2+x^4)dx$$

$$= k\left(81x - 6x^3 + \tfrac{1}{5}x^5\right)\Big|_{0}^{3} = \frac{648}{5}k.$$

Thus $\bar{y} = \dfrac{\frac{648}{5}k}{36k} = \dfrac{648}{5\cdot 36} = \dfrac{18}{5}$,

so the center of mass is at $\left(0, \tfrac{18}{5}\right)$.

FLUID FORCE

6-42

A trough with triangular ends 2 feet wide at the top and 3 feet deep is filled with oil of density 30 pounds per cubic foot. Compute the force exerted by the oil on one end of the tank.

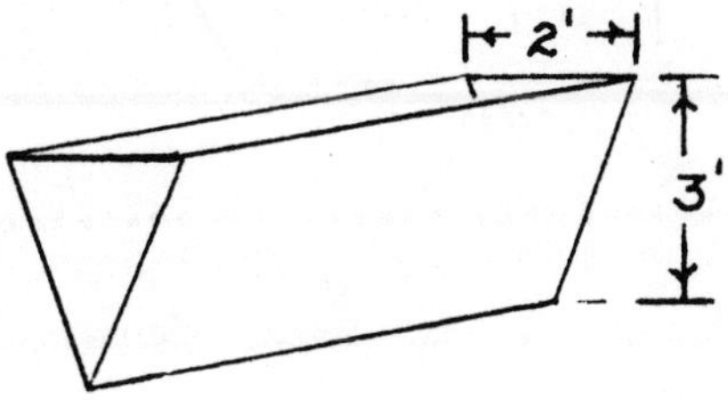

****

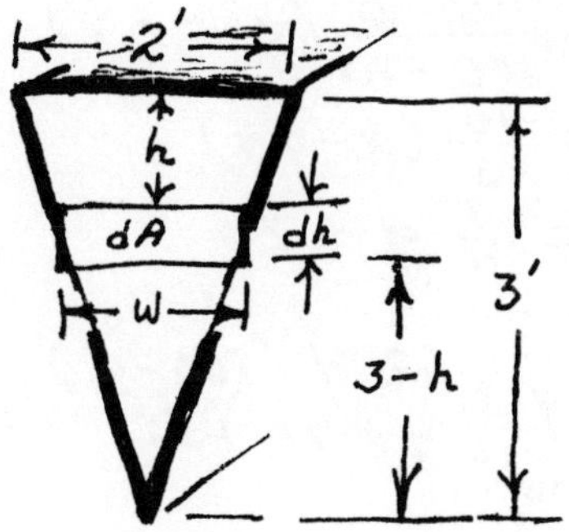

$$\frac{w}{3-h} = \frac{2}{3} \qquad w = \frac{2}{3}(3-h)$$

$$P = \frac{F}{A} = \frac{\delta V}{A} = \delta h \qquad \delta = \frac{30\ \text{lb}}{ft^3}$$

$$F = \int P\, dA \qquad dA = w\, dh$$

$$F = \int_0^3 \delta h\, w\, dh = \int_0^3 30 h \tfrac{2}{3}(3-h)\, dh$$

$$F = 20 \int_0^3 (3h - h^2)\, dh = 20 \left[\frac{3h^2}{2} - \frac{h^3}{3} \right]_0^3 = 10 \cdot 27 - 20 \cdot 9$$

$$= 90 \text{ pounds.}$$

6-43

A dam has a vertical face in the shape of an isosceles trapezoid with the dimensions shown. The surface of the water is at the top of the dam. Find the total force of the water on the face of the dam.

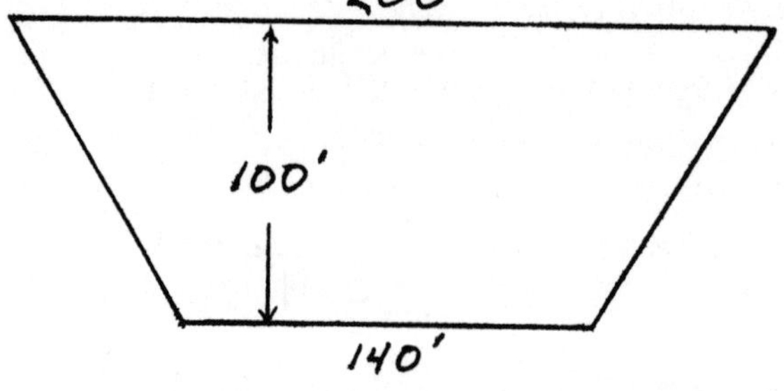

****

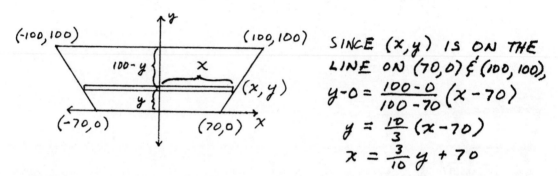

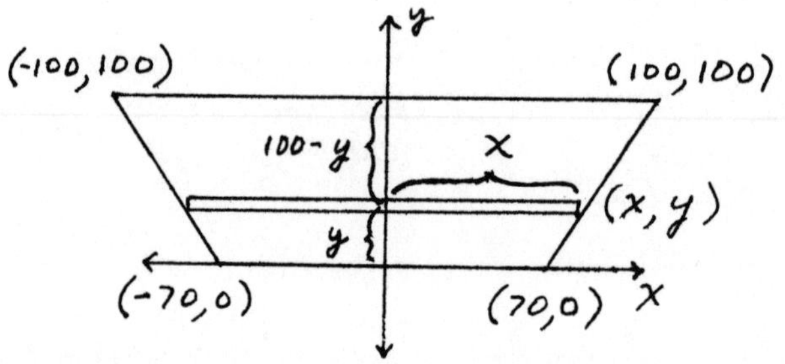

SINCE (x,y) IS ON THE LINE ON $(70,0)$ & $(100,100)$,

$$y-0 = \frac{100-0}{100-70}(x-70)$$

$$y = \frac{10}{3}(x-70)$$

$$x = \frac{3}{10}y + 70$$

PRESSURE (LBS/FT^2) AT DEPTH $100-y$: $62.4(100-y)$

SURFACE AREA (FT^2) OF STRIP: $2\left(\frac{3}{10}y+70\right)dy$

$$\int_0^{100} 62.4(100-y)2\left(\frac{3}{10}y+70\right)dy = 124.8\int_0^{100}\left(-\frac{3}{10}y^2 - 40y + 7000\right)dy$$

$$= 124.8\left(-\frac{1}{10}y^3 - 20y^2 + 7000y\right)\Big|_0^{100}$$

$$= 124.8(-100{,}000 - 200{,}000 + 700{,}000)$$

$$= 124.8\,(400{,}000)$$

$$= 49{,}920{,}000$$

THE TOTAL FORCE IS ABOUT 50 MILLION LBS.

6-44

Mr. Smith bought 500 pigs at a sale. After buying these pigs he did not have enough money to buy a watering trough; hence, he decided to build one from some old lumber. The trough is 10 feet long, 4 feet wide across the top and the ends are equilateral triangles with a vertex pointing down. What is the total force on the end of the trough when it is full of water whose density is 62.4 pounds per cubuc foot?

****

Since the triangle is equilateral, we may assume that it is located as follows:

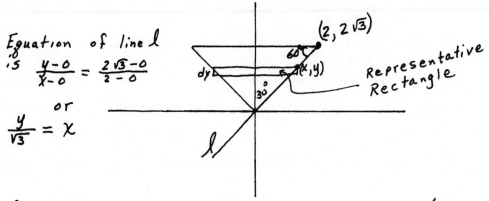

Equation of line l is $\frac{y-0}{x-0} = \frac{2\sqrt{3}-0}{2-0}$

or

$\frac{y}{\sqrt{3}} = x$

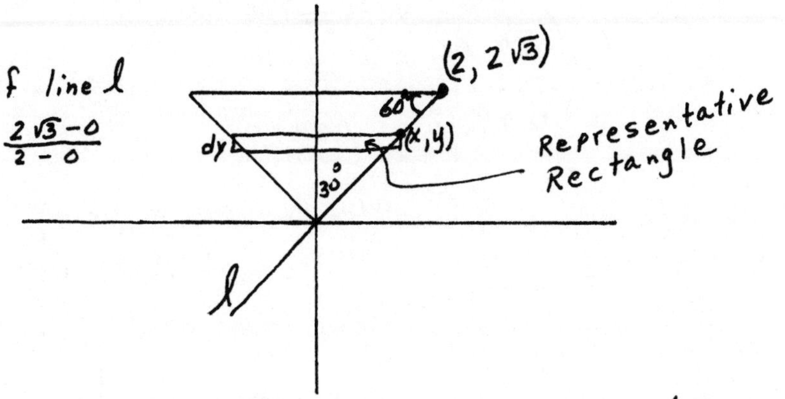

Choose a representative rectangle $(2\sqrt{3} - y)$ units below the surface.

The area of this representative rectangle is $2x\,dy$; hence, the volume of water above this rectangle is

$$(2\sqrt{3} - y)(2x\,dy).$$

So the force on this rectangle is

$$62.4(2\sqrt{3} - y)(2x\,dy).$$

So total force on the end of the trough is given by

$$\text{Total force} = 62.4\int_0^{2\sqrt{3}} (2\sqrt{3} - y)(2x\,dy)$$

or

$$\text{Total force} = 62.4\int_0^{2\sqrt{3}} 2(2\sqrt{3}-y)\left(\frac{y}{\sqrt{3}}\right)dy$$

$$= 124.8\int_0^{2\sqrt{3}}\left(\frac{2\sqrt{3}}{\sqrt{3}}y - \frac{y^2}{\sqrt{3}}\right)dy$$

$$= 124.8\left[y^2 - \frac{y^3}{3\sqrt{3}}\right]_0^{2\sqrt{3}}$$

$$= (124.8)(4)$$

$$= 499.2 \text{ pounds}$$

MISCELLANEOUS

6-45

Suppose that a container is formed by revolving a portion of the curve $y = 4x^4$ about the y-axis.

a) Find the volume of water in the container when it is filled to a depth of h centimeters.

b) If the water is allowed to drain through a small hole in the bottom of any container then Torricelli's Law states that

$$\frac{dV}{dt} = -k\sqrt{h}$$

where V is the volume, k is a constant and h is the depth of the water in the container.

Show that for the container in part (a), $\frac{dh}{dt}$ is constant.

****

a)

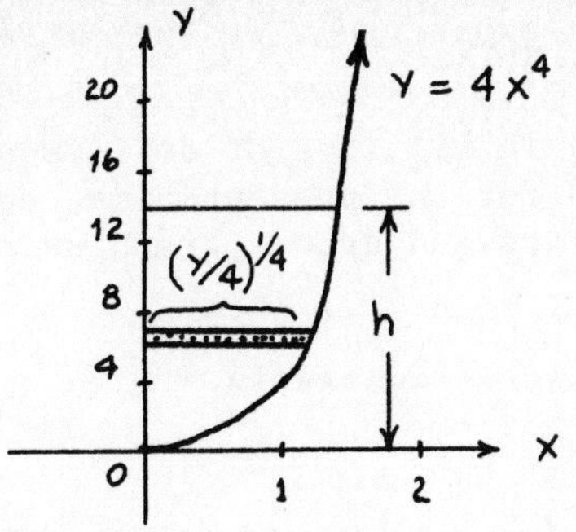

To find the volume of the container consider a horizontal rectangle in the region. As it is revolved about the y-axis a solid disk is formed.

Note that the radius of the disk is found by solving $y = 4x^4$ for x, giving $x = (y/4)^{1/4}$

So, $$V = \int_0^h \pi \left[\left(\frac{y}{4}\right)^{1/4}\right]^2 dy = \frac{\pi}{2}\int_0^h \sqrt{y}\, dy = \frac{\pi h^{3/2}}{3} \text{ cm}^3$$

b) Differentiating the volume obtained in part (a) gives

$$\frac{dV}{dt} = \frac{\pi}{3} \cdot \frac{3}{2} \cdot h^{1/2} \cdot \frac{dh}{dt} = \frac{\pi\sqrt{h}}{2} \cdot \frac{dh}{dt}$$

But by Torricelli's Law, $dV/dt = -k\sqrt{h}$

Thus $$\frac{\pi\sqrt{h}}{2} \cdot \frac{dh}{dt} = -k\sqrt{h} \Rightarrow \frac{dh}{dt} = \frac{-2k}{\pi}$$

And $\frac{dh}{dt}$ is constant, as desired.

6-46

An investment company is advised by its economic expert that its net investment flow should be approximated by the function :

$g(t) = t^{1/3}$, where g(t) is counted in billions of dollars per year and t is number of years. Considering t = 0 for the present time, calculate the capital formation over the next 12 years and draw a graph using values at 4, 8, and 12 years respectively.

****

For the purpose of drawing the graph, we assume that capital at time $t = 0$ is \$ 0.00. The time interval is given as 4, 8, and 12 years. From the given equation, we write:

for 4 years, $\int_0^4 t^{1/3}\,dt$

$$= \frac{3}{4} t^{4/3} \Big/_0^4 = \frac{3}{4}\left[4^{4/3} - 0^{4/3}\right]$$

$= \$\ 4.76$ billion

for 8 years, $\int_0^8 t^{1/3}\,dt$

$$= \frac{3}{4} t^{4/3} \Big/_0^8 = \frac{3}{4}\left[8^{4/3} - 0^{4/3}\right]$$

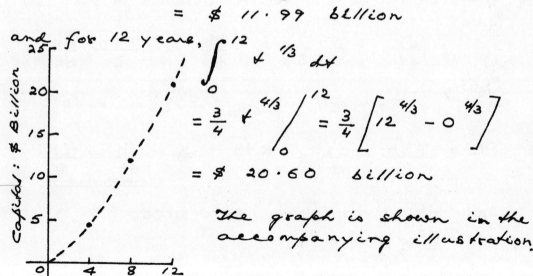

$= \$\ 11.99$ billion

and for 12 years, $\int_0^{12} t^{1/3}\,dt$

$$= \frac{3}{4} t^{4/3} \Big/_0^{12} = \frac{3}{4}\left[12^{4/3} - 0^{4/3}\right]$$

$= \$\ 20.60$ billion

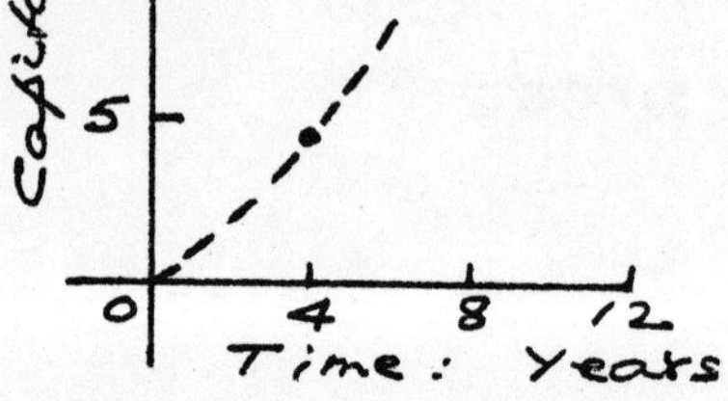

The graph is shown in the accompanying illustration.